Cover Caption:

Heliophysics studies the entire domain of the Sun from its interior structure and dynamics to the interaction of solar wind with the interstellar medium. Its discoveries are relevant to all three of the other branches of the Science Mission Directorate (SMD). The changing solar output interacts with each of the planets in different ways. The Sun is the only star that can be examined in sufficient detail to use it as the Rosetta Stone for stellar astrophysics. Solar flares and coronal mass ejections modulate the geospace environment affecting not only Earth-orbiting satellites including Global Positioning Systems (GPSs), but also perturbing ground-based technologies such as electric power grids. Extreme events could affect the safety of astronauts operating beyond the protective shield of the magnetosphere. Heliophysics works on spatial scales that span over 10 decades and temperatures that range from 200 K to over 20 MK. For details of these images see the key on Page v.

NASA/TM-2009-214178

GSFC Heliophysics Science Division 2008 Science Highlights

Holly R. Gilbert
NASA Goddard Space Flight Center, Greenbelt, Maryland

Keith T. Strong
SP Systems, Inc., Greenbelt, Maryland

Julia L.R. Saba
Lockheed Martin Advanced Technology Center, Palo Alto, California

Elaine R. Firestone and Robert L. Kilgore
TRAX International, Greenbelt, Maryland

National Aeronautics and
Space Administration

Goddard Space Flight Center
Greenbelt, Maryland 20771

March 2009

The NASA STI Program Office ... in Profile

Since its founding, NASA has been dedicated to the advancement of aeronautics and space science. The NASA Scientific and Technical Information (STI) Program Office plays a key part in helping NASA maintain this important role.

The NASA STI Program Office is operated by Langley Research Center, the lead center for NASA's scientific and technical information. The NASA STI Program Office provides access to the NASA STI Database, the largest collection of aeronautical and space science STI in the world. The Program Office is also NASA's institutional mechanism for disseminating the results of its research and development activities. These results are published by NASA in the NASA STI Report Series, which includes the following report types:

- TECHNICAL PUBLICATION. Reports of completed research or a major significant phase of research that present the results of NASA programs and include extensive data or theoretical analysis. Includes compilations of significant scientific and technical data and information deemed to be of continuing reference value. NASA's counterpart of peer-reviewed formal professional papers but has less stringent limitations on manuscript length and extent of graphic presentations.

- TECHNICAL MEMORANDUM. Scientific and technical findings that are preliminary or of specialized interest, e.g., quick release reports, working papers, and bibliographies that contain minimal annotation. Does not contain extensive analysis.

- CONTRACTOR REPORT. Scientific and technical findings by NASA-sponsored contractors and grantees.

- CONFERENCE PUBLICATION. Collected papers from scientific and technical conferences, symposia, seminars, or other meetings sponsored or cosponsored by NASA.

- SPECIAL PUBLICATION. Scientific, technical, or historical information from NASA programs, projects, and mission, often concerned with subjects having substantial public interest.

- TECHNICAL TRANSLATION. English-language translations of foreign scientific and technical material pertinent to NASA's mission.

Specialized services that complement the STI Program Office's diverse offerings include creating custom thesauri, building customized databases, organizing and publishing research results ... even providing videos.

For more information about the NASA STI Program Office, see the following:

- Access the NASA STI Program Home Page at http://www.sti.nasa.gov/STI-homepage.html

- E-mail your question via the Internet to help@sti.nasa.gov

- Fax your question to the NASA Access Help Desk at (443) 757-5803

- Telephone the NASA Access Help Desk at (443) 757-5802

- Write to:
  NASA Access Help Desk
  NASA Center for AeroSpace Information
  7115 Standard Drive
  Hanover, MD 21076–1320

Available from:

NASA Center for AeroSpace Information
7115 Standard Drive
Hanover, MD 21076-1320

National Technical Information Service
5285 Port Royal Road
Springfield, VA 22161

GSFC Heliophysics Science Division 2008 Science Highlights

Table of Contents

Cover Figure Keys v
FOREWORD vii
PREFACE x
INTRODUCTION 1
   The Sun 1
   The Inner Heliosphere 3
   Geospace 3
   The Outer Heliosphere 5
THE HSD ORGANIZATION 6
FACILITIES 8
2008 SCIENTIFIC HIGHLIGHTS 9
   EUNIS Probes a Coronal Bright Point 9
   STEREO Tracks Solar Wind Compression Regions 12
   Coronagraph Prototype Tested at the Chinese Eclipse 13
   ST5 First to Demonstrate Gradiometer Capabilities 14
   Voyager 2 Crosses the Termination Shock 15
HSD PROJECT LEADERSHIP 18
DEVELOPING FUTURE HELIOPHYSICS MISSION CONCEPTS 20
   Introduction 20
HELIOPHYSICS EDUCATION AND PUBLIC OUTREACH 22
   Education 22
   Public Outreach 23
   General Outreach Activities 24
   Sun–Earth Connection Education Forum 24
   Public Media 25
SCIENCE INFORMATION SYSTEMS 26
COMMUNITY COORDINATED MODELING CENTER 29
   Science Support 30
   Space Weather Modeling 31
TECHNOLOGY DEVELOPMENT 33
APPENDIX 1: INDIVIDUAL SCIENTIFIC RESEARCH 35
APPENDIX 2: HSD PUBLICATIONS AND PRESENTATIONS 104
   Journal Articles 104
   Submitted / In Press 125
   Presentations 131
APPENDIX 3: CURRENT HSD MISSIONS 146
   Interstellar Boundary Explorer (IBEX) 146
   Communications/Navigation Outage Forecasting System (C/NOFS) 147
   Aeronomy of Ice in the Mesosphere (AIM) 149
   Time History of Events and Macroscale Interactions During Substorms
      (THEMIS) 150
   Solar Terrestrial Relations Observatory (STEREO) 151
   Hinode 152
   Solar Radiation and Climate Experiment (SORCE) 154

Ramaty High Energy Solar Spectroscopic Imager (RHESSI) 155
Thermosphere, Ionosphere, Mesosphere, Energetics, and Dynamics (TIMED) 156
Cluster 157
Two Wide-Angle Imaging Neutral-Atom Spectrometers (TWINS) 158
Transition Region And Coronal Explorer (TRACE) 159
Fast Auroral Snapshot Explorer (FAST) 160
Polar 161
Advanced Composition Explorer (ACE) 162
Solar and Heliospheric Observatory (SOHO) 163
Wind 164
Geotail 165
Ulysses 166
Voyager 167

APPENDIX 4: FUTURE MISSIONS 168
Solar Dynamics Observatory (SDO) 168
Radiation Belt Storm Probes (RBSP) 169
Magnetospheric MultiScale (MMS) 170
Solar Orbiter 171
Solar Probe Plus 172
Sentinels 173
Solar C 174
Geospace Electrodynamic Connections (GEC) 175
Magnetospheric Constellation (MagCon) 176
Interstellar Probe 177
Stellar Imager (SI) 178

APPENDIX 5: ACRONYM LIST 179

GSFC Heliophysics Science Division 2008 Science Highlights

Cover Figure Keys

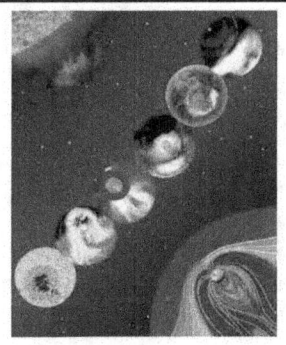	The solar system is dominated by the Sun. Its constantly changing output of energy, particles, and magnetic fields affects the entire solar system in a variety of fascinating ways. The branch of science devoted to understanding these phenomena and their relevance to other scientific disciplines is called "heliophysics." It covers vast ranges of temperature, densities, spatial scales, and energies. Particles and fields from the Sun impact the Earth causing the geomagnetic storms and the buildup of particles in the radiation belts. The effects of solar variability form an intricate chain of complex physical processes that stretch to the edge of interstellar space, defining the heliosphere—the volume of space in which humans will exist and operate for the foreseeable future.
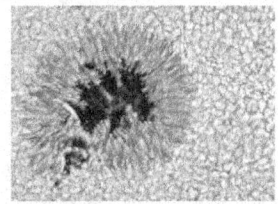	Sunspots are areas of concentrated strong magnetic field, often larger that the Earth. They appear dark because they inhibit the flow of energy to the surface of the Sun and thus are cooler. They appear in cycles peaking approximately every 11 years. At the onset of a cycle, they appear at high latitudes, but emerge at successively lower latitudes until most of the remaining activity is concentrated near the equator at solar minimum.
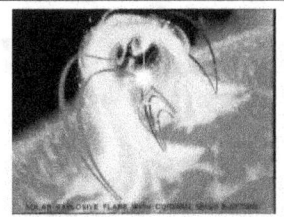	Strong fields on the Sun can become twisted and stretched, storing energy that can be released in flares. These explosive events can increase short wavelength emissions from the Sun by many orders of magnitude in seconds, accelerate particles to near relativistic velocities, and heat the solar atmosphere to over 20 MK. The effects range from direct atmospheric heating, shortening the life of Low-Earth Orbit (LEO) satellites, to complete radio blackouts by saturating those bands, GPS for example.
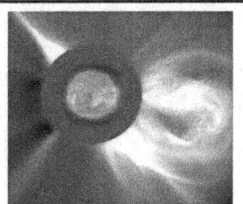	Another type of eruptive event involving the reconfiguration and reconnection of magnetic fields is the Coronal Mass Ejection. They are vast expulsions of coronal plasma into interplanetary space where the fast moving (sometimes >2000 km/s) ejecta interact with the solar wind. Sometimes they strike Earth's magnetosphere and, if the magnetic configuration is favorable, can cause geomagnetic storms and other space weather effects.
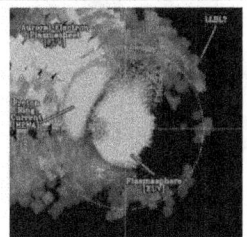	Superposition of Exteme Ultraviolet (EUV) imaging of the cold plasma in the inner magnetosphere (10,000 K), the hot plasma electrons (10,000,000 K) precipitating in the aurora, the hot plasma ions of the ring current (up to 1,000,000,000 K). These last make up the hottest region in the inner solar system, even hotter than the solar corona (1,000,000 K).

GSFC Heliophysics Science Division 2008 Science Highlights

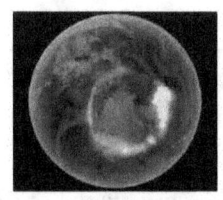	Auroras are one of the most obvious consequences of a geomagnetic storm in progress. Particles accelerated by magnetic reconnection in the tail of Earth's magnetosphere streaming along field lines impact the upper atmosphere creating these dazzling and dynamic light shows in the polar regions.
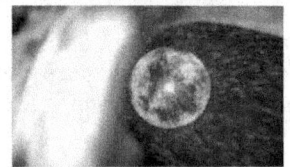	The heliosphere represents the entire domain of the Sun stretching over 100 AU in the ram direction and many times that towards its tail. This shape is rather like Earth's magnetosphere in the solar wind. Here, the solar wind slows due to the inflow of interstellar gases forming a shock region. Similarly, the interstellar material is slowed as it encounters the solar wind building up a bow shock.
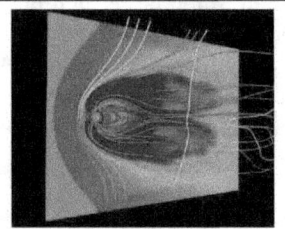	A model of the complex interactions of the Earth's magnetic field with the solar wind using a 3-D modeling code developed and tested at the GSFC Community Coordinated Modeling Center (CCMC) facility. It is one of many models that are contributed to the center to help merge the different codes into a unified "Sun to mud" model of the Sun–Earth system, and to transition those models to operational use by the space weather community.
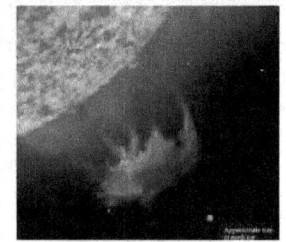	A solar prominence seen in He II 304A—relatively cold, dense material that was originally suspended in the 1-million degree corona is ejected from the solar surface at very high velocity.
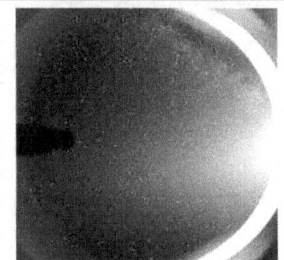	Although this is not apparent by just looking at the cover figure, the background is the zodiacal light against a star background taken by the Heliospheric Imager (HI) instrument on the Solar Terrestrial Relations Observatory (STEREO).

FOREWORD

This report is intended to record and communicate to our colleagues, stakeholders, and the public at large about heliophysics scientific and flight program achievements and milestones for 2008, for which NASA Goddard Space Flight Center's Heliophysics Science Division (HSD) made important contributions. HSD comprises approximately 261 scientists, technologists, and administrative personnel dedicated to the goal of advancing our knowledge and understanding of the Sun and the wide variety of domains that its variability influences.

- GSFC has a key role in every major NASA heliophysics mission currently flying or being planned.
- HSD made some amazing scientific contributions to better understanding the Sun and its sphere of influence in 2008.
- HSD has supported the development of the mission concepts and technologies vital to the future of heliophysics.

HSD's Mission is to explore the Sun's interior and atmosphere, discover the origins of its temporal variability, understand its influence over the Earth and the other planets, and determine the nature of the interaction between the heliosphere and the local interstellar medium.

Activities:
- Lead science investigations involving flight hardware, theory, and data analysis and modeling that will answer the strategic questions posed in the Heliophysics Roadmap;
- Lead the development of new solar and space physics mission concepts and support their implementation as Project Scientists;
- Provide access to measurements from the Heliophysics Great Observatory through our Science Information Systems;
- Communicate science results to the public and inspire the next generation of scientists and explorers.

Outcomes:
- Open the frontier to space environment prediction;
- Understand the nature of our home in space;
- Safeguard the journey of exploration.

The HSD scientists and staff have had a highly productive year. The "Extreme Ultra-Violet Normal Incidence Spectrograph (EUNIS)," Principal Investigator (PI) Doug Rabin (671), and "Twin Rockets to Investigate Cusp Electrodynamics (TRICE)," GSFC Lead Co-Investigator (Co-I) R. Pfaff (674), had fully successful flights on 2007 November 16 and December 10. The United States Air Force Communications/Navigation Outage Forecast System (C/NOFS) and NASA's Interstellar Boundary Explorer (IBEX) mission were launched on 2008 April 17 and October 19, respectively. Rob Pfaff (674) is the PI for the Vector Electric Fields Instrument (VEFI) science investigation on C/NOFS, which includes five instruments, including contributions by several of HSD partner institutions.

Tom Moore (670.0) and his plasma instrumentation team provided key portions of the IBEX low-energy neutral-atom imager. With Mike Kaiser (674) as Project Scientist, the two STEREO spacecraft reached their operational separations in 2008 and are now observing the start of the rising phase of Solar Cycle 24. Joe Davila (671) led the development of the inner coronagraph (COR1) on STEREO and is Goddard's lead Co-I on the Sun–Earth Connection Coronal and Heliospheric Imager (SECCHI) coronal transient imaging investigation.

The past year also saw HSD scientists producing new scientific and technical results. Len Burlaga (673) and his colleagues on the GSFC magnetometer team—M. Acuna (695), J. Connerney (695) and R. Lepping (674)—observed Voyager 2's crossing of the termination shock caused by the collision of the solar wind with the local interstellar medium at the outer edges of the solar system. In contrast to the earlier Voyager 1 encounter with this important boundary, the Voyager 2 measurements revealed a new type of collisionless shock with strong coupling to interstellar neutral particles. Jim Slavin (670.0), Guan Le (674), and Y. Wang (UMBC/GEST) published the first gradiometric measurement of auroral field-aligned currents in low-Earth orbit using the Space Technology 5 (ST5) micro-satellites. This approach provides current density measurements that are not aliased by plasma waves and other temporal variations, and it demonstrates the capabilities of micro-satellites to produce unique, research-grade observations.

HSD published 213 papers in refereed journals in 2008, of which 44% had a GSFC scientist as first author; 70 more papers are either in press or submitted. In addition, HSD scientists gave 162 presentations at 65 different conferences, most of which were given at the two American Geophysical Union (AGU) meetings. Appendix 2 details the body of work that the group has accomplished in the last year or so.

HSD scientists garnered many top professional honors in 2008. The EUNIS rocket team won the Robert H. Goddard Exceptional Achievement Award for Science. The National Academy of Sciences awarded the 2008 Arctowski Medal to Leonard Burlaga. Spiro Antiochos was elected a fellow of the American Physical Society. Jim Slavin received the NASA Exceptional Achievement Medal for Space Technology 5, NASA's micro-satellite constellation technology mission launched in 2006. Anand Bhatia was invited to give the Professor S.C. Sircar Memorial Lecture at the Indian Institute for Cultivation of Science in Kolkta, India, which is one of India's highest scientific honors. The Family Science Night team won the Robert H. Goddard Exceptional Achievement Award for Outreach. The TIMED[*] and STEREO teams won NASA Group Achievement Awards.

---

[*] TIMED: Thermosphere-Ionosphere-Mesosphere Energetics and Dynamics

This Annual Report includes an overview of our current organization and facilities, FY08 science highlights, our programs and missions, and our education and outreach accomplishments for FY08. It will also be made available online. Please check our Web site for details (http://hsd.gsfc.nasa.gov/). We thank you for you interest in our program, and would appreciate any feedback via the Web site.

—James A. Slavin, Director
NASA GSFC Heliophysics Science Division

PREFACE

This document summarizes the work performed in FY08 by members of the Heliophysics Science Division (HSD) who conducted research, developed models, designed and built instruments, managed projects, and carried out numerous other activities that have produced significant contributions to understanding the domain of the Sun. Only a fraction of these can be even briefly highlighted in this report.

The production of this report was guided by Holly Gilbert, HSD Associate Director for Science, who, together with Keith Strong (SP Systems), assembled the contributions and checked the report for accuracy, made suggestions regarding its content, and contributed to several sections. Julia Saba (Lockheed Martin) and Yvonne Strong (American Society of Microbiology) helped with the scientific and technical editing, and Julie Rickman (GSFC) contributed by compiling the large bibliography of the HSD. The publication was led by Elaine Firestone (TRAX International, from Goddard's Technical Information and Management Services Branch, [TIMS], Code 271) and graphics support was provided by Robert Kilgore (TRAX International and TIMS). Many others in HSD helped with useful and constructive suggestions concerning the organization of this report and its content.

—Holly Gilbert and Keith Strong
March 2009

> HSD's goal is to further scientific understanding of the nature and origins of the ever-changing physical conditions throughout the vast volume of space controlled by the Sun, and in which humanity will function for the foreseeable future.

INTRODUCTION

Heliophysics is simply the study of the domain of the Sun—the heliosphere—from its nuclear core where hydrogen is transmuted into helium, producing the energy that drives changes throughout the entire solar system, to the edge of interplanetary space where the solar wind and magnetic fields cede control of the local physical conditions to the interstellar medium. That represents over 10^8 AU^3 suffused with outflowing plasma, magnetic fields, and solar radiation, with temperatures ranging from over 20 MK to near absolute zero.

The heliosphere is an interconnected network of physical processes driven by the relentless, but varying, outflow of energy from the Sun in the form of electromagnetic radiation from γ-rays to radio emissions, ionized and neutral particles, and magnetic fields. All these forms of solar emissions interact in different ways in the wide range of environments from the hot solar interior, through the Sun's thin surface layers, into the extended solar corona, and throughout interplanetary space to the very edge of the solar system. Along their tortuous path, these emissions interact with different planetary environments, comets, asteroids, and interstellar gas—each with its individual response to the changing solar stimuli. HSD's goal is to understand this system of systems.

Accomplishing this goal involves the study of the complex interactions between electromagnetic radiation, plasmas, and magnetic fields, with three principal objectives:

- *To understand the changing flow of energy and matter throughout the Sun, solar atmosphere, heliosphere, and planetary environments*
- *To explore the fundamental physical processes that characterize space plasmas*
- *To define both the origins and the societal impacts of variability in the Sun–Earth System*

There are four major physical domains that encompass HSD's mission: the Sun, the inner heliosphere, geospace, and the outer heliosphere.

The Sun

The Sun generates not only energy, but also magnetic fields. The solar dynamo, combined with both radial and latitudinal differential rotation, generates magnetic fields and stores vast amounts of energy in them. Convection in the outer layers of the Sun and the natural buoyancy of the flux ropes drag these strong fields (>1000 G) to the surface, as evidenced by the presence of sunspots and faculae, which change the solar spectral irradiance that provides the energy to drive Earth's weather and climate system. Thus a significant change in solar irradiance could affect Earth's climate.

Energy is transported from the solar core region (the inner 20% of the solar radius) by photon radiation out to about 70% of the solar radius, but then convectional transport takes over, carrying most of the energy to the surface of the Sun—the photosphere—where the optical depth of the solar plasma drops, so that much of the energy can be radiated away into space. The falling temperature gradient as the energy flows outward seems well understood; however, just above the relatively cooler surface layers, the temperature of the plasma rises rapidly again to form a 1-MK corona. The physical processes involved in creating and maintaining the corona are not completely understood as yet.

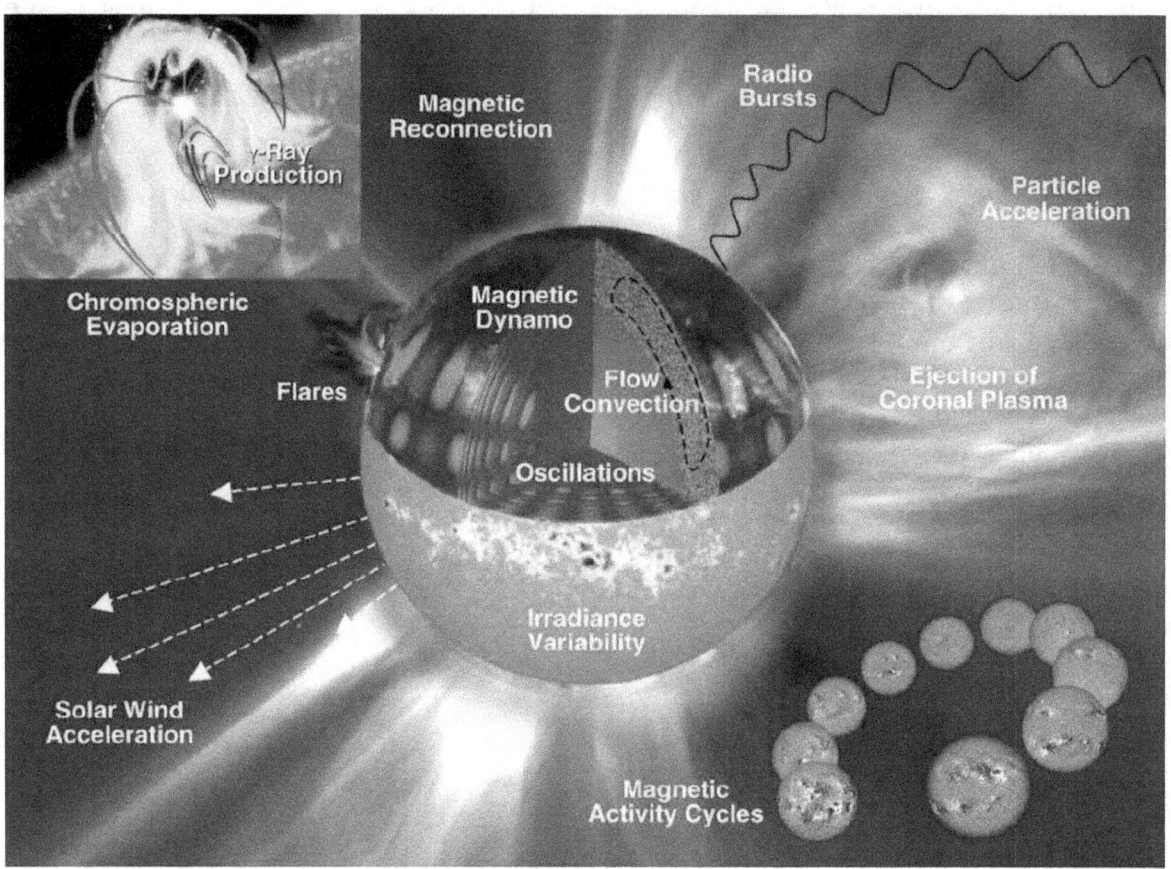

Anatomy of the Sun.

The energy stored in the solar magnetic fields is often released suddenly by magnetic reconnection to produce flares and coronal mass ejections (CMEs). Flares produce emissions from γ-rays to radio wavelengths, accelerate solar energetic particles, and transport material from the lower layers of the atmosphere up into the hot corona via chromospheric evaporation, accompanied by ejection of material away from the Sun. CMEs are vast ejection events that can grow to be many times the size of the Sun and at times move with velocities exceeding 2000 km/s.

Another type of mass outflow from the Sun is more continuous but also highly variable: the solar wind, which flows out along the spiraling solar magnetic field with velocities of between 300 and 700 km/s at various temperatures, densities, and compositions. The manner of its acceleration is still not completely understood.

The Inner Heliosphere

This is the region between the Sun and Jupiter that is filled with outflowing, supersonic solar wind and frozen-in spiraling magnetic fields—the Parker Spiral. The streams of solar plasma evolve significantly as they pass through this region, where fast streams of solar wind plough into slower-moving ones forming shocks. Transients, such as CMEs, reshape the ambient environment. Some CMEs move faster than the local solar wind, building up high-density fronts that form shocks where particles are accelerated to extremely high energies. CMEs can expand as they move outwards, leaving low-density regions behind the propagating front.

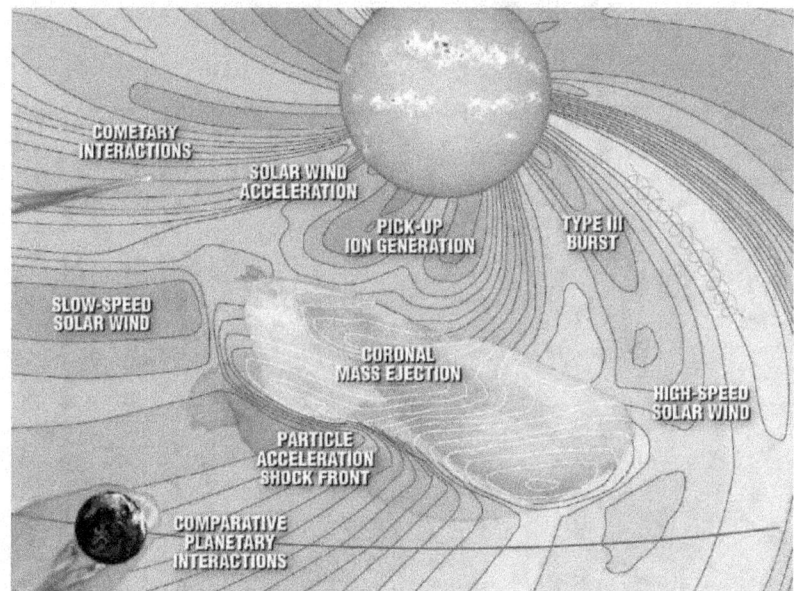

A CME propagates through the heliosphere towards Earth.

Electrons flow along the large-scale magnetic field lines, producing radio bursts of various types and show which field lines remain connected back to the Sun and which ones have reconnected.

Because of the spiral nature of the fields, Earth is better connected to the west limb of the Sun; thus, an event at, or near, the west limb of the Sun is more likely to be geoeffective than one in the eastern hemisphere of the Sun. Photons take only 8 min to arrive at Earth from the Sun, and high-energy protons can be detected a few minutes later, whereas material from a CME event seen on the Sun may take up to 3 days to arrive.

Geospace

Earth's magnetic field acts as a barrier to most of the harmful particle fluxes originating from the Sun. Much of the solar wind is directed around the magnetosphere, which forms a teardrop-shaped shield around Earth. The shape and size of the magnetosphere change as solar wind conditions vary. Earth's magnetic field is compressed within about 10 Earth radii (R_E) on the sunward side of the planet and stretched out by many tens of Earth radii on the anti-sunward side. The faster and denser the winds, the more the fields are compressed and stretched. The direction of the interplanetary field profoundly influences the effects seen; a CME with an oppositely directed magnetic field will more likely reconnect, allowing more energetic particles to enter geospace, and producing a more energetic geomagnetic storm.

Where the solar wind and Earth's magnetic field collide there is a bow shock region. The fields become weak and disorganized as they interact in the magnetosheath. The surface where Earth's magnetic pressure is balanced by the solar wind is called the magnetopause; it often ripples and flaps in the solar wind, and parts constantly magnetically reconnect and break away. In the polar region, there is a path where the solar particles have easy access to Earth's magnetosphere—this is the cusp region.

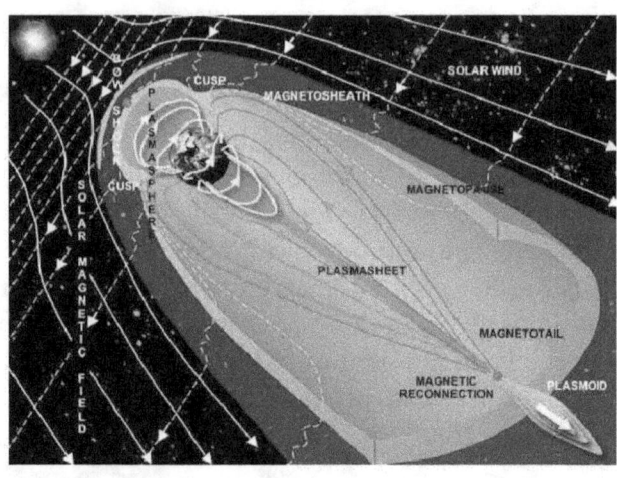

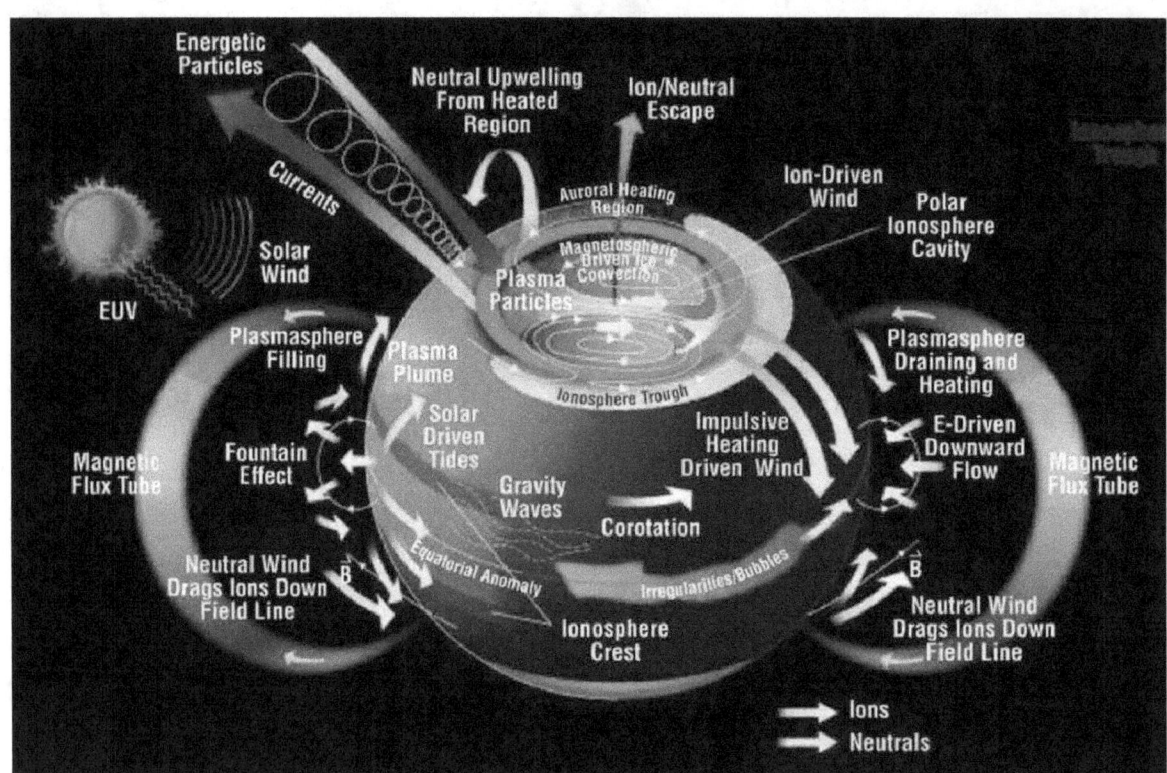

Some of the complex physical processes that occur when incoming radiation and charged particles interact with Earth's dynamic atmosphere.

Closer to Earth, a completely different set of closely coupled processes occur. Here, solar radiation has one of its primary effects. Ultraviolet (UV) light from the Sun, especially when enhanced by flare emissions, heats the neutral atmosphere, increasing its scale height, which reduces the orbital life of spacecraft. Gravity waves propagate up from below in the neutral atmosphere. The thermosphere becomes partly ionized by absorbed solar emissions that heat it increasingly with altitude. As a result of changing inputs, it becomes highly dynamic. The ionosphere, the first completely ionized layer of Earth's atmosphere, becomes very disturbed at times of high solar activity; these disturbances, in turn, interfere with many forms of HF radio communications, and cause bulk outflow of the ionosphere into the magnetosphere from the auroral zones. High-energy electrons and protons become trapped in the Van Allen radiation belts. The outer belt gathers electrons from the aftermath of geomagnetic storms and substorms, accumulating enough plasma pressure to produce a ring-shaped current around Earth that substantially inflates the geomagnetic field. The inner belt originates from cosmic rays interacting with the upper atmosphere. Charged particles spiral along the field lines, mirroring as they descend into the stronger polar fields.

The Outer Heliosphere

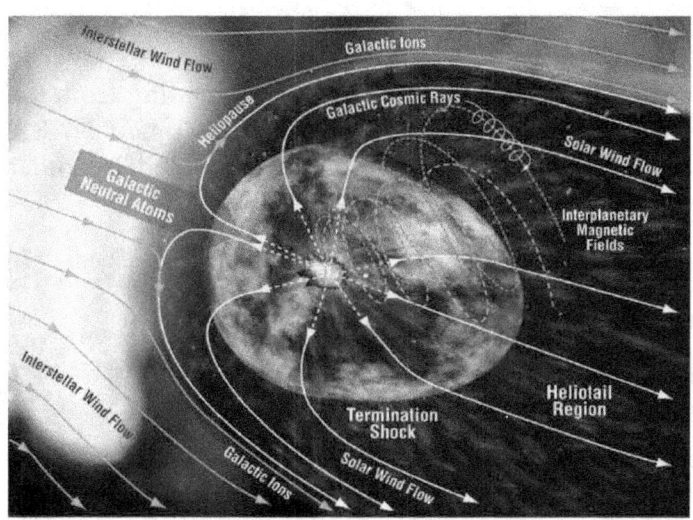

The characteristics of the heliosphere significantly change past the orbit of Jupiter—this area is called the outer heliosphere. Here, the solar wind and transients interact with the gas-giant planets, which are very different from their rocky cousins in the inner heliosphere.

In this region, the nature of the outflowing plasma also changes; the interplanetary magnetic field becomes mostly azimuthal, and therefore, is perpendicular to the solar wind flow and is where solar transients and interplanetary shocks catch up with each other and form large Global Merged Interaction Regions. In addition, a larger portion of the solar wind is composed of photoionized interstellar neutral particles, known as pickup ions.

With increasing distance from the Sun, the particle and field pressure of the solar wind decreases until it reaches pressure balance with the local interstellar medium. This boundary is called the heliopause; however, before reaching this boundary, the solar wind has to slow abruptly below its supersonic speed at the termination shock, and start deflecting toward the heliotail and continue slowing down in the region known as the heliosheath. The interaction of the heliosphere with the interstellar medium is analogous to the solar wind deflecting around Earth's magnetosphere. It is postulated that the interstellar plasma could also flow at supersonic speeds, necessitating the existence of an external bow shock and a pileup of interstellar particles upstream of the heliosphere, known as the hydrogen wall. Besides low-energy neutral particles, the extremely high-energy galactic cosmic rays also enter the heliosphere, but not without first being modulated by the periodically varying heliospheric magnetic fields. The termination shock and heliosheath are thought to also be the source of an extra, anomalous component of the cosmic rays observed at Earth.

> HSD is an internationally recognized research organization dedicated to the furtherance of scientific understanding of all aspects of heliophysics.

THE HSD ORGANIZATION

At the end of FY08, HSD was composed of 261 scientists, engineers, and other staff supported by a small management and administrative team. The Division employs 75 civil servants (29%), as well as university scientists working under cooperative agreements (13%), technical contractor staff (41%), NASA Postdoctoral Fellows (6%), emeritus scientists (6%), and other support staff (5%).

The organization is divided into four Laboratories, each with its own Chief, as shown in the figure below. HSD is supported primarily by competitively awarded funding from the NASA Science Mission Directorate with the remainder made up of assigned NASA tasks, Center research and development investments, and funding provided by other Federal agencies.

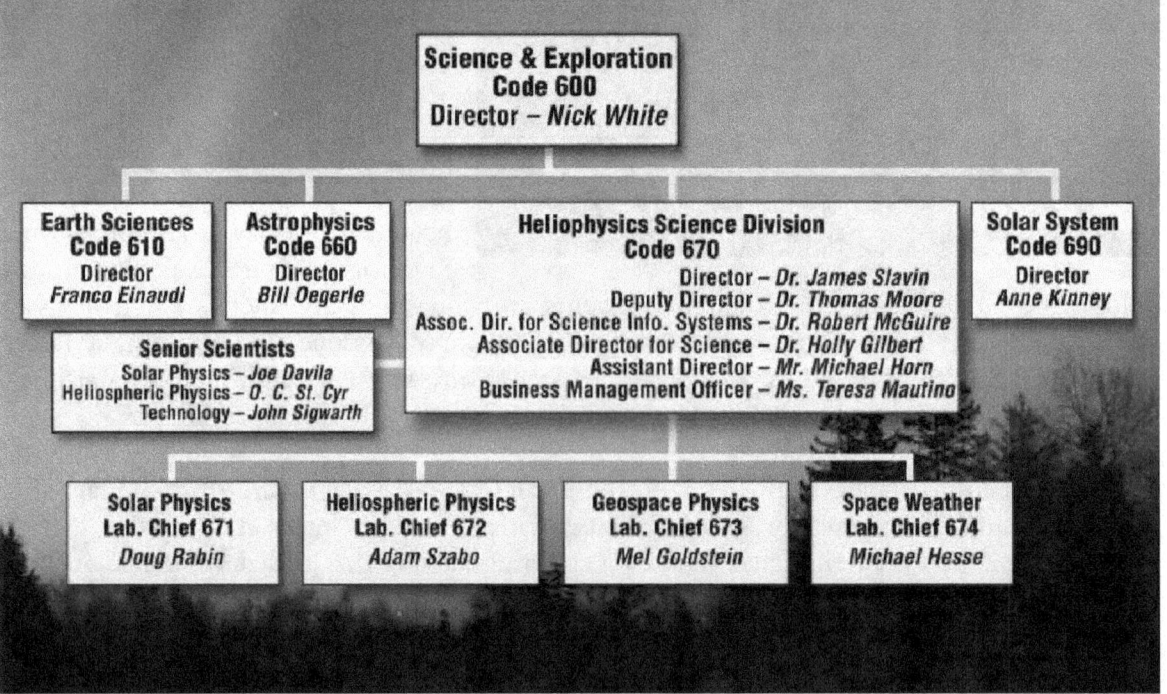

The responsibilities of HSD include:
- Scientific Research: HSD working as Principal Investigators (PIs), Co-Investigators (Co-Is), Instrument Scientists, and flight team members have published 213 papers and given over 185 presentations at scientific meetings in 2008 (see Appendices 1 and 2 for details).
- Project and Mission Scientist Assignments: HSD provides the project and mission scientists who manage operating heliophysics missions, as well as those missions in development (see the Missions sections, Appendices 3 and 4, for details).

- Future mission concept development: HSD provides scientific leadership and technical support for science mission concept development and formulation.
- Data and Modeling Centers: HSD Scientists lead and operate four major centers that provide critical data services and simulation and modeling services to the Heliophysics community. They are the Space Physics Data Facility, the Solar Data Analysis Center, Heliophysics Data and Modeling Consortium, and the Community Coordinated Modeling Center. These centers are funded directly by NASA and reviewed periodically by NASA Headquarters-appointed external committees.
- Education and Public Outreach (E/PO): HSD scientists, led by the Associate Director for Science (Holly Gilbert), carry out a variety of E/PO tasks supported by Project and competitively awarded funding.

Like many Government, academic, and industrial research laboratories that perform basic and applied research in specialized areas, HSD has experienced recruitment challenges in replacing retirees and recruiting new staff to achieve its research goals, in addition to its flight project, and data and modeling Center responsibilities. For this reason, HSD is actively recruiting within the university community to attract new post-doctoral, cooperative agreement, and civil service scientists. For example, in FY08, HSD welcomed 15 postdoctoral fellows to the HSD organization. The Division also has other student development programs to attract more young researchers to space science and retain the most promising candidates. Another important goal is to achieve greater ethnic and gender diversity in the HSD workforce. Over the past year, HSD lost seven staff members to retirement and hired six new civil servants (Holly Gilbert, Spiro Antiochos, Jim Klimchuk, Eric Christian, Judy Karpen, and George Khazanov).

> HSD's goal is to provide a safe and efficient work environment, consistent with being a center of excellence for heliophysical research.

FACILITIES

Building 21 Entrance

HSD has people located in several buildings on the GSFC campus, primarily Buildings 21 and 26. Following the opening of Building 34, plans are in place to consolidate the whole Division into Building 21 over the next two years. Once this is completed, all HSD personnel, offices, laboratories, and other facilities will be housed in the same building for the first time.

In 2008, the renovation of Building 21 was started. The most significant project was the expansion and rehabilitation of the Space Plasma Instrument Facility. The original laboratory configuration included a 753-square feet (ft^2) clean room and 935 ft^2 of laboratory space, and housed an electron vacuum chamber and an ion vacuum chamber. Once the expansion is completed in mid-2009, the Plasma Laboratory will have 1200 ft^2 of clean room, 1000 ft^2 of laboratory space, and an additional electron chamber. Other renovation work included updating a conference room and some initial office upgrades. A display area was added to the lobby to showcase HSD's heritage, along with current missions and scientific results.

> With a strong scientific base, HSD can function effectively to support and promote a vital heliophysics program at NASA

2008 SCIENTIFIC HIGHLIGHTS

While it would not be practical to feature, in detail, all the scientific accomplishments of the HSD team in this type of report, a few outstanding examples of work that was successfully completed in 2008 are presented in this section. A short summary of individual scientific contributions is given in Appendix 1, and more extensive list of publications and presentations can be found in Appendix 2.

EUNIS Probes a Coronal Bright Point

The Extreme-Ultraviolet Normal-Incidence Spectrograph (EUNIS) is a Goddard-designed and operated sounding-rocket instrument that obtains imaged high-resolution solar spectra. It has now had two successful flights, on 2006 April 12 and 2007 November 16, providing spectra of unprecedented temporal resolution (<10 s), as well underflight calibrations for a number of orbiting solar experiments on both occasions (including the SOHO[*]/Coronal Diagnostic Spectrometer and the Hinode/EUV Imaging Spectrometer).

Doug Rabin (PI) and the EUNIS field team at White Sands readying EUNIS for launch

EUNIS gathers spectra in the 170–205 Å and 300–370 Å wavelength regions and is calibrated to an absolute photometric accuracy of <15%. During its 2006 flight, it obtained spectra of a coronal bright point (BP).

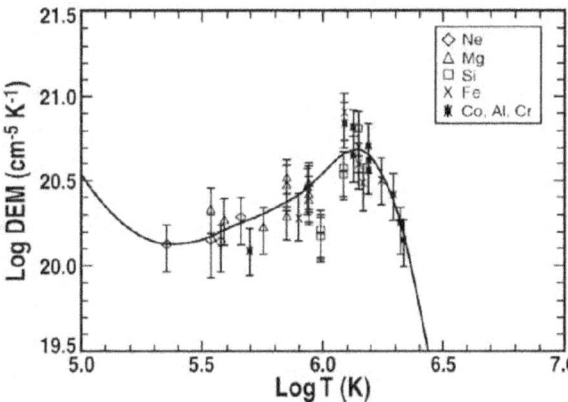

DEM of a coronal bright point.

The quiet solar corona is heated predominantly by small-scale magnetic reconnection events. It is estimated that BPs could contribute 20% or more of this heating.

BPs, first identified from X-ray images, are small (~2–5 Mm) magnetic bipoles that exist for periods from a few hours to a couple of days. There are estimated to be ~1000 on the Sun at any given time. To quantify their role in coronal heating, scientists need to be able to derive the typical physical conditions

---

[*] SOHO: Solar and Heliospheric Observatory

within a BP. The EUNIS data provide that opportunity because of their extensive roster of temperature and density-sensitive line ratios and radiometric precision.

The subsequent analysis enabled the EUNIS team to:

- Derive the first differential emission measure (DEM) of a coronal bright point (see figure),
- Show that BPs have photospheric, not coronal, abundances, and so are likely filled by chromospheric evaporation,
- Conclude that BPs are formed by magnetic-reconnection-driven chromospheric evaporation.

EUNIS is funded for a third flight in 2010, which will explore a different wavelength region with an advanced variable line space grating and more sensitive detectors.

Coronal Tomography

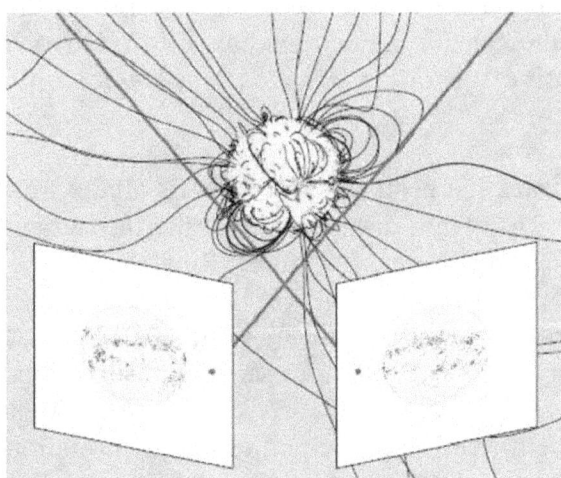

The STEREO COR1 instruments provide white-light observations of the solar corona from 1.5–4 solar radii (R_S) where the transition from closed to open coronal structures takes place. COR1 polarized brightness images from the two spacecraft are used to reconstruct the electron density structure in this region. The reconstruction is performed by the regularized tomography inversion method. This method allows reconstruction only of structures that are stationary during an observation period, i.e., from a quarter to half a solar rotation for periods when the Sun is relative quiet. The reconstructed density structure of the equatorial streamer belt in most of the reconstructed domain is consistent with the variation of the current sheet derived from a potential magnetic field model. This code is being modified to provide these models routinely to help assess the value of tomographic reconstruction in understanding coronal and heliospherical processes.

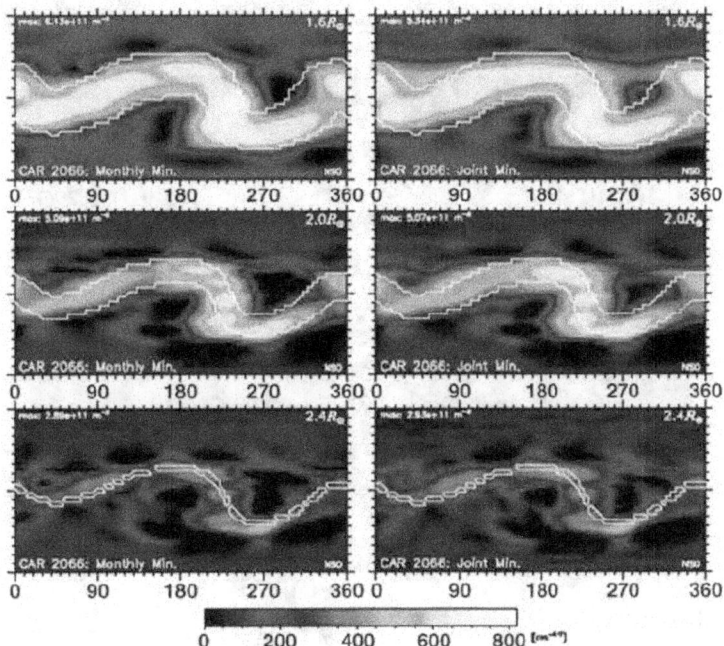

Spherical cross-sections of the reconstructed electron density for Carrington rotation 2066 at heliocentric distances 1.6, 2.0, and 2.4, R_S (the distances are shown in the right upper corners). Left panel corresponds to the reconstruction based on data with monthly minimum background subtraction, and right panel corresponds to the reconstruction based on the same data with joint minimum background subtraction. The white contour lines are boundaries between closed and open magnetic filed lines for potential field approximation based on NSO/GONG data.

STEREO Tracks Solar Wind Compression Regions

The Solar Terrestrial Relations Observatory (STEREO) comprises two spacecraft in the same 1 Astronomical Unit (AU) orbit about the Sun, with one leading Earth and one trailing. They are slowly increasing their relative separation to provide a stereoscopic view of solar phenomena, especially CMEs.

STEREO carries a wide-angle camera that can track the propagation of CMEs all the way from near the Sun to Earth. Moreover, the STEREO team discovered that the instrument can track high-density solar wind features as they propagate out from the Sun.

Joint STEREO and Wind *in situ* observations connected for the first time solar wind compression regions, observed at 1 AU for many years, with remote-sensing observations of density enhancements in the inner heliosphere. This allows the tracking of these solar wind structures back towards the Sun and the identification of the point of their formation when fast streams ram into slower ones. This research is carried out by A. Szabo (672) and B. Thompson (671) in collaboration with S. Plunkett of the National Research Laboratory (NRL).

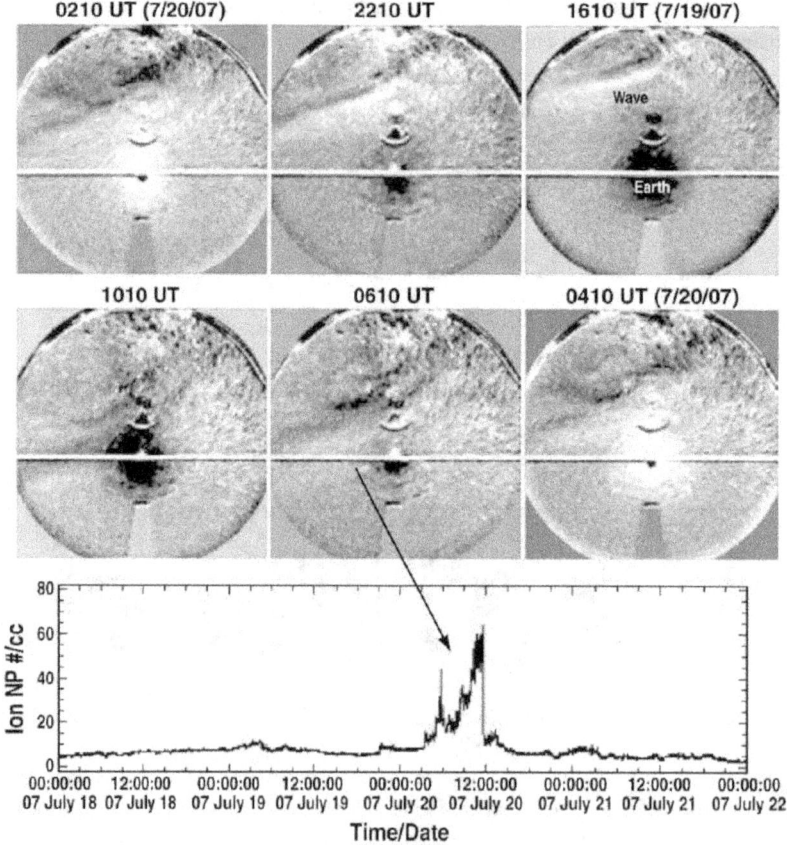

Time-ordered difference images of the heliosphere taken from STEREO, looking towards Earth, with the Sun just off the pictures to the left. The bright band moving across the field of view is the density enhancement. The plot shows the solar wind proton density enhancement observed by Wind/SWE at the same time when the bright HI2 feature reaches Earth.

Coronagraph Prototype Tested at the Chinese Eclipse

On 2008 August 1, a total eclipse of the Sun tracked across Canada, Greenland, Russia, Mongolia, and China. A GSFC group used this opportunity to successfully test two new solar instruments that are candidates for future space science missions.

The Imaging Spectrograph of Coronal Electrons (ISCORE) obtained images in four narrowband filters that will yield maps of electron temperature and flow speed using a method proposed in 1973 and proven in previous GSFC eclipse expeditions using a fiber-fed spectrograph. The filter-based instrument is compact and ideally suited to spacecraft application. A second instrument obtained images with high angular resolution and high cadence (6 frames per second) to capture the dynamics of the coronal structures close to the solar limb (see lower figure). Results from the ISCORE experiment have been submitted for publication.

The eclipse was also an E/PO success. There were live Web casts, educational activities that were made available online, and Web resources detailing the course and causes of the eclipse that were made available to the public. In addition, several HSD scientists were interviewed by national news organizations.

Right to left: Joe Davila (PI), Nelson Reginald, and Doug Rabin in the Gobi desert preparing their instruments for observing the eclipse.

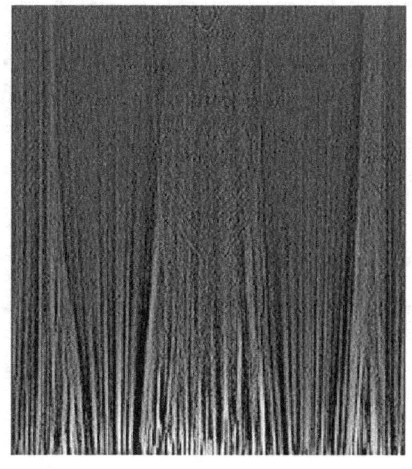

The fine structure of coronal streamers imaged during the eclipse expedition.

Holly Gilbert and Dean Pesnell explain the eclipse on NBC national news.

ST5 First to Demonstrate Gradiometer Capabilities

On 2006 March 22, Space Technology 5 (ST5) was launched into a full-Sun polar orbit (300 × 4500 km) as part of the New Millennium Program. ST5 is NASA's pathfinder for future missions requiring highly capable, affordable, small spacecraft; miniaturized subsystems; and constellation mission operations. ST5 was designed to show the utility of small satellites by flying a 3-microsatellite "string-of-pearls" constellation with miniaturized Fluxgate Magnetometers (FGM) aboard. The mission was designed, built, and operated by GSFC.

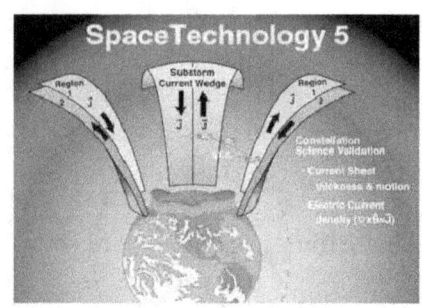

Although the mission ended after 90 days of operation, the recent ST5 data analysis has demonstrated the first magnetic field gradiometric measurement capabilities:

o First magnetic field gradiometer measurements of field-aligned current intensity

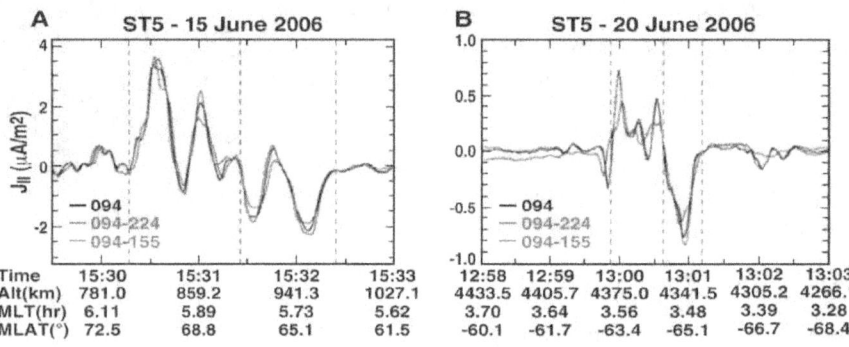

o First crustal magnetic anomaly measurements using two-point magnetic field gradiometer.

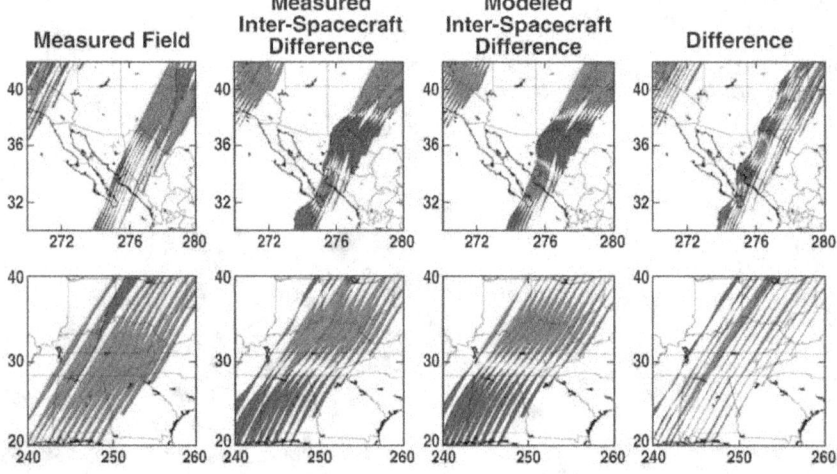

Voyager 2 Crosses the Termination Shock

The solar wind plasma is ejected by the Sun in all directions at supersonic speeds. It continues to travel at these speeds until it encounters the interstellar medium, made up of plasma, neutral gas, and dust. This decelerates the solar wind and causes a shock front to form—the termination shock, which effectively marks the outermost limit of the Sun's sphere of influence. The distance of this boundary from the Sun varies depending on the direction of the wind relative to the motion of the Sun around the galaxy (the ram direction), the density of the local interstellar medium, the strength and direction of the solar magnetic fields, and the velocity and density of the solar wind.

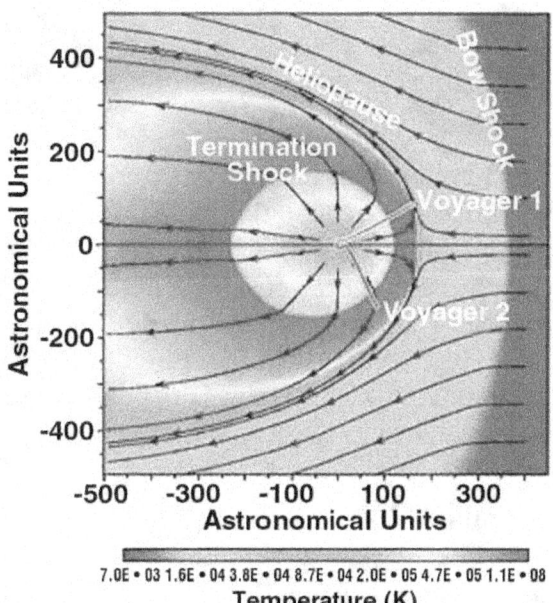

Structure of the heliosphere, showing the ellipsoidal termination shock and heliopause, and the trajectories of Voyager 1 and Voyager 2.

The two Voyager spacecraft have been heading out of the solar system for over 30 years. Voyager 1 crossed the termination shock in December 2004 at 94 AU; however, Voyager 1 does not have a functioning solar wind plasma detector. Therefore, detailed observation of the termination shock had to wait till Voyager 2 crossed it in September 2007. Since then, Voyager 2 has remained in the heliosheath, the region of shocked solar wind. An unexpected finding is that the flow there is still supersonic with respect to the thermal ions downstream of the termination shock. According to the latest data, the wind downstream of the shock is cooler and faster moving than researchers had anticipated. One interpretation is that the solar wind is imparting energy to neutral atoms from the interstellar gas and causing them to ionize.

The detailed structure of the neutral pickup ion-dominated shock has been extensively studied by the Voyager magnetometer team of L. Burlaga (673), M. Acuna (695), and N. Ness (from the University of Delaware).

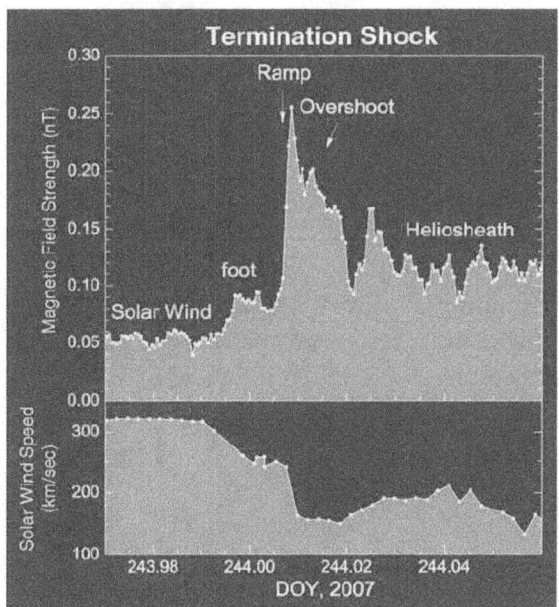

Observations of the termination shock by Voyager 2, showing the magnetic structure (top plot) and the decreasing speed across the termination shock (bottom plot).

International Heliophysical Year

The International Heliophysical Year (IHY) is a global activity to stimulate heliophysical activities throughout the world. Planning began in 1999, and activities begin in 2001, culminating in the "year" spanning 2007–2009 [the 50th anniversary of the International Geophysical Year (IGY)]. There were IHY national committees in nearly 90 nations involving the efforts of thousands of scientists. In March 2007 the IHY was officially opened, and what a success it has been!

A total of 65 science campaigns were initiated for the IHY, including the "Whole Heliosphere Interval," which involved over 100 scientists and 50 instruments. IHY science sessions were held worldwide, and many IHY science meetings helped open the frontier of heliophysical research. Perhaps the highlight of these was the 2007 "IHY-Africa Space Weather and Education Workshop" in Addis Ababa, Ethiopia, which brought together over 100 scientists from around the world, including IHY leaders from African nations, to stimulate space science research throughout Africa.

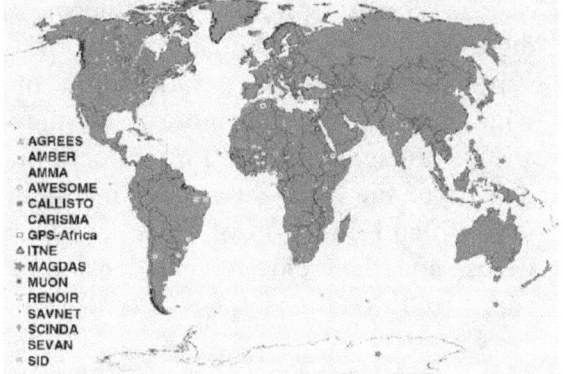

Map indicating instrument deployments supported as part of the IHY/UNBSS program. Please note that the map does not include the majority of the observatories participating in IHY, only those participating in the instrument development program.

A major achievement of the IHY has been the deployment of instrument arrays, particularly in regions where measurements were most needed. Each program involved a team of scientists working together on problems of mutual interest. The distributed instrument program, developed through the United Nations Basic Space Science (UNBSS) program, facilitated the deployment of a wide range of instrument arrays all over the world, including places that were thought to be unreachable (see figure).

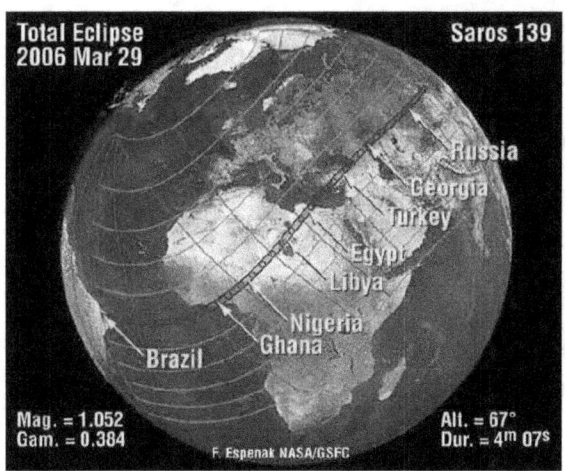

IHY Eclipse Viewing Stations along the path of totality allowed people the opportunity to safely observe an eclipse, receive educational materials, and discuss the event with researchers.

Of course, outreach and education activities were essential for the development of heliophysics worldwide. There were traveling exhibitions, summer schools, museum partnerships, student-run experiments, online activities, and multi-lingual materials. Excellent examples of these were activities centered around natural phenomena that attract a great deal of public interest.

An IGY History program was also developed, which recognized individual participants, gathered historical material, and held IGY history events and celebrations worldwide.

GSFC Role: GSFC scientists perform the majority of the duties of the secretariat. The IHY Executive Director is Joe Davila, the International Coordinator is Nat Gopalswamy, and the Director of Operations is Barbara Thompson.

Status: IHY's official closing ceremony will transpire at the United Nations in Vienna on 2009 February 18. Transitional activities, which are aimed towards the solidification and continuance of the IHY legacy, will continue through 2011.

> Heliophysics missions consist of an intricate network of space observatories that monitor the Sun's variability and its effects on the entire solar system, life, and society.

HSD PROJECT LEADERSHIP

Heliophysics covers a vast volume of space from the center of the Sun to the edge of interstellar space. HSD looks at diverse physical processes from magnetic reconnection to particle acceleration, shock formation to convection, thermal conduction to nonthermal heating. This requires the use of many different observing techniques from a variety of venues. In recent years, it has become apparent that little further progress will be made unless these phenomena are studied as a "system of systems."

A fleet of spacecraft, often referred to as the Heliophysics Great Observatory (HGO), are currently being flown; each spacecraft is addressing a different aspect of the problems that HSD is trying to unravel. Goddard's HSD is the nexus of these projects, having a management role in all but three of these missions and scientific participation in all of them. See the table below.

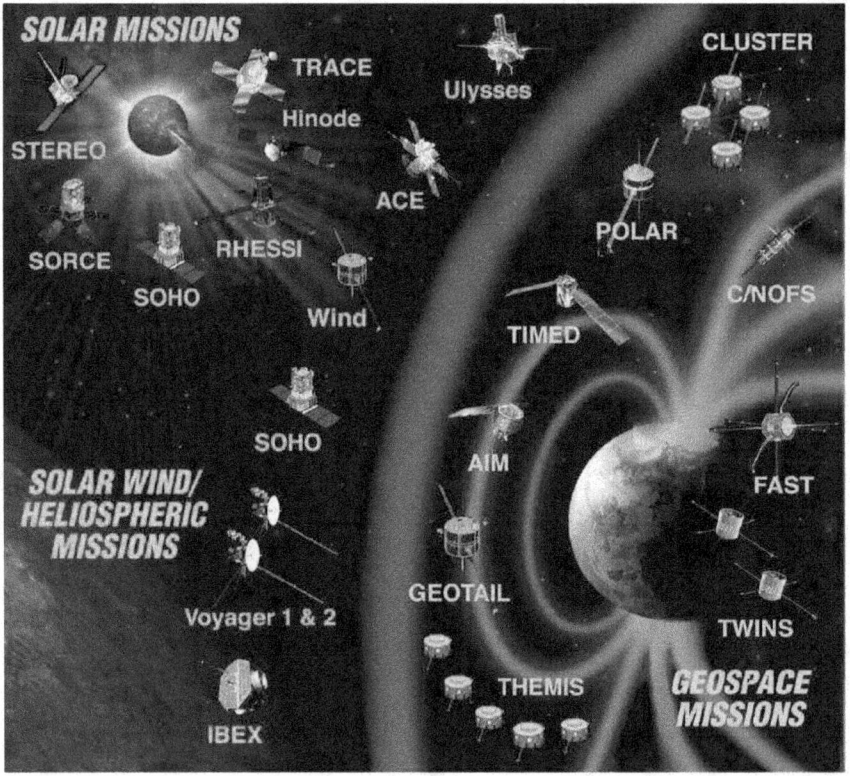

The Heliophysics Great Observatory: Currently, Heliophysics has three missions observing the ionosphere, mesosphere, and thermosphere regions; six missions observing the magnetosphere and ionosphere; seven missions sampling the solar wind and heliosphere; and six missions observing the Sun. These make up a powerful combination of in situ *and remote-sensing instruments.*

The heliophysics missions flying in FY08:

Solar Terrestrial Probes		Senior Project Scientist: Jim Slavin (GSFC)	
Living With a Star		Senior Project Scientist: Chris St. Cyr (GSFC)	
Mission	**Launch Date**	**Project/Mission Scientist**	**Scientific studies of:**
IBEX	2008 Oct 19	Bob MacDowall (GSFC)	Outer heliosphere
C/NOFS	2008-Apr-16	Rob Pfaff (GSFC)	Ionospheric scintillations
AIM*	2007-Apr-02	Hans Mayr (GSFC)	Mesospheric clouds
THEMIS†	2007-Feb-17	Dave Sibeck (GSFC)	Substorms
STEREO	2006-Oct-25	Mike Kaiser (GSFC)	Coronal mass ejections
Hinode	2006-Sep-22	John Davis (MSFC)	Solar magnetic fields
SORCE‡	2003-Jan-25	Robert Callahan (GSFC)	Solar irradiance
RHESSI	2002-Feb-05	Brian Dennis (GSFC)	High-energy flares
TIMED	2001-Dec-07	Dick Goldberg (GSFC)	Thermosphere, ionosphere, mesosphere
Cluster	2000-Jul-16 2000-Aug-09	Mel Goldstein (GSFC)	Particle / magnetic field interactions
TWINS§	2000-Mar-25	Mei-Ching Fok (GSFC)	Magnetosphere
TRACE**	1998-Apr-01	Joe Gurman (GSFC)	Solar atmosphere
ACE††	1997-Aug-25	Tycho von Rosenvinge (GSFC)	Solar wind
FAST‡‡	1996-Aug-21	Rob Pfaff (GSFC)	Aurora Acceleration
Polar	1996-Feb-24	John Sigwarth (GSFC)	High latitude magnetosphere
SOHO	1995-Dec-02	Joe Gurman (GSFC)	Solar interior, magnetic activity cycle, corona, solar wind
Wind	1994-Nov-01	Adam Szabo (GSFC)	Solar wind
Geotail	1992-Jul-24	Guan Le (GSFC)	Dynamics of the magnetotail
Ulysses	1990-Oct-6	Ed Smith (JPL)	Solar wind
Voyager 1 Voyager 2	1977-Sep-05 1977-Aug-20	Ed Stone (JPL)	Heliosphere

Appendix 3 is a brief outline (in reverse chronological order) of the purpose and status of each of the current heliophysics missions and GSFC's role in them.

---

* AIM: Aeronomy of Ice in the Mesosphere
† THEMIS: Time History of Events and Macroscale Interactions during Substorms
‡ SORCE: Solar Radiation and Climate Experiment
§ TWINS: Two Wide-Angle Imaging Neutral-Atom Spectrometers
** TRACE: Transition Region and Coronal Explorer
†† ACE: Advanced Composition Explorer
‡‡ FAST: Fast Auroral Snapshot Explorer

> "The best present is a future"—the next-generation instruments and spacecraft will gather data from new vantage points, creating a vibrant future for heliophysics.

DEVELOPING FUTURE HELIOPHYSICS MISSION CONCEPTS

Introduction

At the moment, there are a number of future heliophysics missions in various stages of development. Key to the future of the Heliophysics Great Observatory (HGO) and the strength of this discipline are vitality of its strategic vision and the pipeline of new space missions that are required to realize the vision. These missions always build on past scientific achievements and technical capabilities, but it is essential that they produce more than incremental results and that some of the more challenging missions be undertaken not in spite of, but rather because of, the challenges they present to heliophysics and NASA. All of these missions are part of the Solar Terrestrial Probe (STP) or the Living With a Star (LWS) program. Heliophysics is a partner, with Astrophysics, in the Explorer program (e.g., TRACE and RHESSI) and expects a steady stream of grass-roots-developed, external Principal Investigator led Explorer missions that will achieve more focused, nearer-term science objectives. There are also mission-of-opportunity payloads on other Government and commercial spacecraft, and collaborations involving international missions (e.g., Hinode and SOHO). There are also NASA space technology development programs and the New Millennium program (e.g., ST5), but these are beyond the scope of this document.

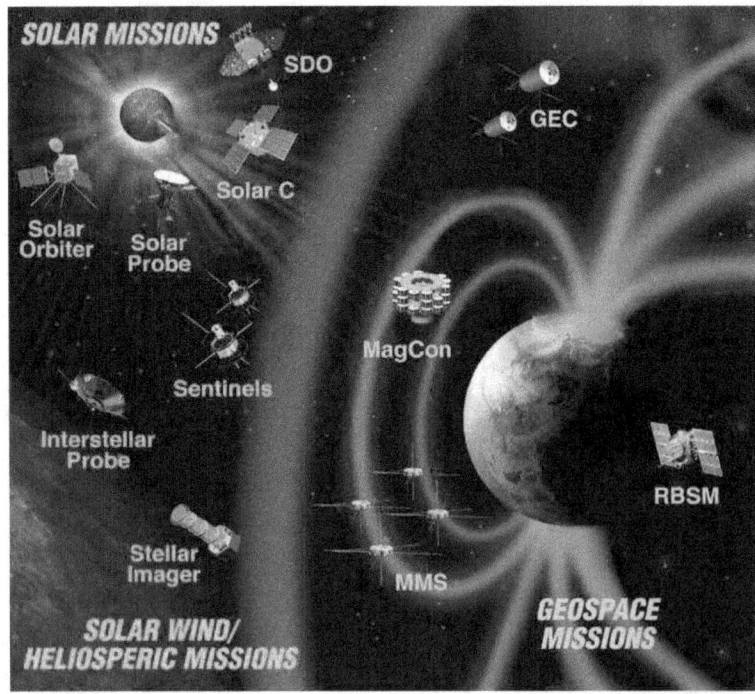

The next generation HGO elements. Two missions will focus on the Ionosphere, Mesophere, and Thermosphere regions; three missions will solve key questions regarding magnetic reconnection and charged particle acceleration in the magnetosphere; four missions will measure the solar wind, determine its acceleration mechanism in the corona, and evolution as it moves out into the heliosphere; and four missions will probe the outer layers of the Sun, its atmosphere, and the eruption of flares and coronal mass ejections. The missions marked in yellow are currently just concepts that are being developed and are not yet in formulation.

The heliophysics missions planned to fly beyond FY08:

Mission	Planned Launch	Project/Mission Scientist	Scientific Studies of:
SDO*	2009	Dean Pesnell (GSFC)	Solar magnetic fields, corona
RBSP†/ITSP‡	2012	Barry Mauk (APL)	Earth's radiation belts
MMS§	2013	Mel Goldstein (GSFC)	Magnetosphere
Solar Orbiter	2017	Chris St. Cyr (GSFC)	Hi-res solar imaging
Solar Probe+	2018	Robert Decker (APL)	Origin of the solar wind
Sentinels	2017	Adam Szabo (GSFC)	Solar Wind / CMEs
Solar C	Concept	TBD	High solar latitudes imaging or atmospheric spectroscopy
GEC**	Concept	TBD	Earth's electrodynamic environment
MMC††	Concept	TBD	Global magnetosphere processes
Interstellar Probe	Concept	TBD	Outer heliosphere
Stellar Imager	Concept	TBD	Activity cycles of Sun-like stars

The nature and timing of the missions may well change as a result of the NASA Heliophysics Roadmap study and the National Research Council (NRC) decadal survey. Appendix 4 provides some of the details of each of these missions.

---

* SDO: Solar Dynamics Observatory
† RBSP: Radiation Belt Storm Probes
‡ ITSP: Ionosphere-Trosposphere Storm Probes
§ MMS Magnetospheric Multiscale
** GEC: Geospace Electrodynamics Connections
†† MMC: Magnetosphere Mission Concept

> Using NASA missions and activities to inspire, engage, educate, and employ is a key component in creating a vibrant future for heliophysics.

HELIOPHYSICS EDUCATION AND PUBLIC OUTREACH

Significant opportunities exist to extend the impact of heliophysics science and related mission activities to engage and inspire students in formal education settings, audiences at informal learning centers, and the general public across the nation via the press and other communication outlets. The complexity of heliophysics requires a systems approach, and our E/PO efforts must mirror this challenge. In FY08, the Division has taken steps to focus and invigorate its outreach activities by creating an E/PO leadership position—Associate Director for Science—filled by Dr. Holly Gilbert. Under the lead of Dr. Gilbert, the Heliophysics E/PO will show an increase in coordination in the development and evaluation of HSD's programs and products. Heliophysics outreach goals are to:

o Communicate heliophysics scientific discoveries and advances
o Raise public understanding of space weather impacts
o Inspire the next generation of American space scientists

These goals can be reached by strategically integrating components of Education (K–12, higher education, and informal education) and Public Outreach (e.g., media). Strong university relationships also play a key role in HSD's success, including adjunct support, faculty visitors, and interactions created during sabbaticals spent at university.

Education

Summer Programs

Through many different summer programs offered (which will be integrated into one large summer program starting in FY09), heliophysicists supported over 11 summer interns (high school and undergraduate) in addition to several graduate students through the Graduate Student Summer Program.

Co-op

The Cooperative Education Program is an important link in the educational process that integrates college level academic study with full-time meaningful work experience. This is achieved through a working agreement between GSFC and a number of educational institutions. HSD supported three Co-op students in FY08.

Postgraduate Programs

o **GSRP:** The Graduate Student Researchers Program provides qualified graduate students, in residence at their home institutions, with fellowship support on research projects of mutual interest to the student and the GSFC mentor. HSD currently supports seven GSRP students.
o **NPP:** The NASA Postdoctoral Program (NPP) provides talented postdoctoral scientists and engineers with valuable opportunities to engage in ongoing NASA research programs. HSD currently has 15 NPP staff members.

Public Outreach

Given the large number of missions in the heliophysics fleet, all of the mission-related outreach activities in a report of this scope cannot be listed, but a few illustrative examples are provided:

- **Community Events at Professional Science Meetings**: The SDO team, in partnership with the Rochester Institute of Technology Center for Imaging Science Insight Lab and with support from the American Astronomical Society (AAS) and the American Geophysical Union (AGU) has implemented two new E/PO programs held prior to each of the AAS and AGU meetings. The Educator Reception is an opportunity for local teachers to interact one-on-one with traveling scientists. During the reception, the teachers can obtain materials to take back to the classroom to excite their students in current science. AstroZone and Exploration Station are each a 4-hour open house for local families, aimed at making science more accessible to the general public and increasing public awareness of the current science being performed. The first Exploration Station was held at the joint AGU/SPD* meeting in Ft. Lauderdale, and the third AstroZone was held at the AAS meeting in St. Louis. These two programs will now each be held twice a year. Emilie Drobnes (SDO E/PO Lead) presented a paper on the Exploration Station and AstroZone programs at the international Committee on Space Research (COSPAR) meeting in July in Montreal.

- **Data Distribution:** as part of the SOHO/STEREO Pick of the Week feature, STEREO images and movies are often sent out to over 230 museums and science centers through ViewSpace kiosks and the American Museum of Natural History's AstroBulletins. Handouts and 3-D movies from STEREO were showcased at Goddard Day in Annapolis, Maryland; Goddard's LaunchFest; NASA Night at the Air and Space Museum; the American Association of Retired Persons (AARP) Convention; and the Association of Science Technology Centers conference. High-resolution STEREO footage was provided to several production companies for displays and TV productions.

- **Distance Learning Network:** the Endeavour Program, a collaboration between STEREO/NASA and selected middle-school students from 18 school systems in northeastern Pennsylvania, offered student teams learning opportunities with a NASA solar mission. This involved receiving a distance learning (DLN) overview of STEREO and 3-D imagery. The student teams conduct research on the mission and space weather, make their own 3-D images, and present their images and explain how STEREO is advancing solar science, also via DLN. A similar program engaged middle school students in several other Pennsylvania school districts in 2008.

- **Web:** SOHO Explore is the SOHO education and outreach Web site, including an introduction to "Our Star the Sun" with images and videos, activities, and lesson plans using SOHO data, educational materials, and an "Ask Dr. SOHO" feature.

---

* SPD: Solar Physics Division of the AAS

General Outreach Activities
Family Science Night
Family Science Night is a 2-hour monthly program for middle school students and their families designed to change their perceptions of science and how they respond to science. The FY09 season will serve as final testing of the program, facilitator guides, and evaluation tools. HSD hopes to begin a first round of dissemination next summer. The Family Science Night team, led by SDO E/PO Lead Emily Drobnes, received the Robert H. Goddard award at the September 10th ceremony.

Sunday Experiment

The Sunday Experiment is a 4-hour open house for the DC metro area. During the event, visitors participate in various hands-on activities set up throughout the museum at the GSFC Visitor Center. The goal of this program is to expose the local community to GSFC and to begin to change their perceptions of science and scientists. It is hoped that this program can serve as a hook into other GSFC programs, such as Family Science Night and the model rocket event, and start to build a GSFC E/PO pipeline of sorts.

Museum Alliance: The Smithsonian Folklife Festival

The 42nd Annual Smithsonian Folklife Festival was held 2008 June 25–July 6 on the National Mall in Washington, D.C. The Heliophysics Division was well represented as several scientists and E/PO experts volunteered for the 10-day event that attracted over one million festival attendees. The solar telescope was the hit of the space science area. Other hands-on materials, such as the 3-D compasses for mapping out magnetic fields, were also very popular.

LaunchFest 2008
Thousands converged on GSFC in September 2008 to celebrate LaunchFest, where exhibits and activities demonstrated Goddard's many current and future missions. Heliophysics activities, specifically the "Walk on the Sun" exhibit about STEREO and the STEREO exhibit showing 3-D animations of the Sun, were extremely popular.

Sun–Earth Connection Education Forum
The Sun–Earth Connection Education Forum (SECEF) is a partnership between GSFC and University of California Berkeley's Space Sciences Laboratory. SECEF brings together the rich expertise of scientists, educators, and museums to develop innovative products and programs with the goal to increase science literacy and focus attention on

the active Sun and its effects on Earth. An example of a successful SECEF activity is Sun–Earth Day:

Sun–Earth Day is an annual national event (usually in March) supported by the SECEF, Sun-Solar-System-Connection scientists, and spacecraft missions. The goal is to share the science of the Sun with educators, students, and the general public via informal learning centers, the Web, television, etc., through high-profile, well-supported annual events. For Sun–Earth Day 2008, SECEF involved a worldwide audience in the celebration of the IHY, "Space Weather Around the World." Mission science and mission-developed resources were highlighted on the Web site and in the 15,000 packets that were distributed widely and used for training in educator workshops. Through podcasts, a quarter of a million individuals were reached. On 2008 August 1, the Sun–Earth Day team, with support from GSFC's Public Affairs Office and NASA TV, co-produced with the Exploratorium a Web cast of the total solar eclipse from China.

Public Media

GSFC's outstanding data visualization team—the Scientific Visualization Studio (SVS)—produces superior quality images, animations, and data visualizations in support of a wide range of heliophysics communications and science activities, including NASA Public Affairs press releases, live presentations, various print publications, television, and video documentaries. In FY08, new animations and visualizations were created to describe mission science, for example, from IBEX, THEMIS, and SDO, allowing the complete, complex, heliophysics story to be told in a way that is understandable to the public.

Other successes of FY08 include over 50 stories (in press releases, www.nasa.gov, and science.nasa.gov), written by HSD's Public Affairs writing team, which was made possible by a significant investment in heliophysics writing support. NASA heliophysics has a direct conduit to spaceweather.com audiences through support of writer Tony Phillips. This coordination is allowing HSD to make greater impact and obtain greater attention for heliophysics science results.

In FY08, two live-shot campaigns (live TV scientist interviews with TV stations around the country) were produced, one based on auroras and incorporating the THEMIS mission, and the other based on the 2008 eclipse. Additionally, for the first time in 2008 HSD produced vodcasts (i.e., video podcasts) for YouTube, iTunes, and www.nasa.gov for some of the major heliophysics stories. The vodcast for the 2008 eclipse got 100,000 views on YouTube, and an eclipse preview featuring SDO had 80,000 page views. THEMIS also benefited from vodcasts. The Associated Press picked up the vodcast titled, "THEMIS: Spring is Aurora Season" resulting in coverage on almost every news site:

http://www.nasa.gov/centers/goddard/news/topstory/2008/aurora_live.html.

> The congruence of expertise and experience in HSD makes it a natural center for leadership in the design and implementation of science information systems.

SCIENCE INFORMATION SYSTEMS

NASA builds and flies heliophysics spacecraft and instruments to collect observational data and increase understanding of the heliophysical system and its detailed processes as can be pictured in a classic research wheel.

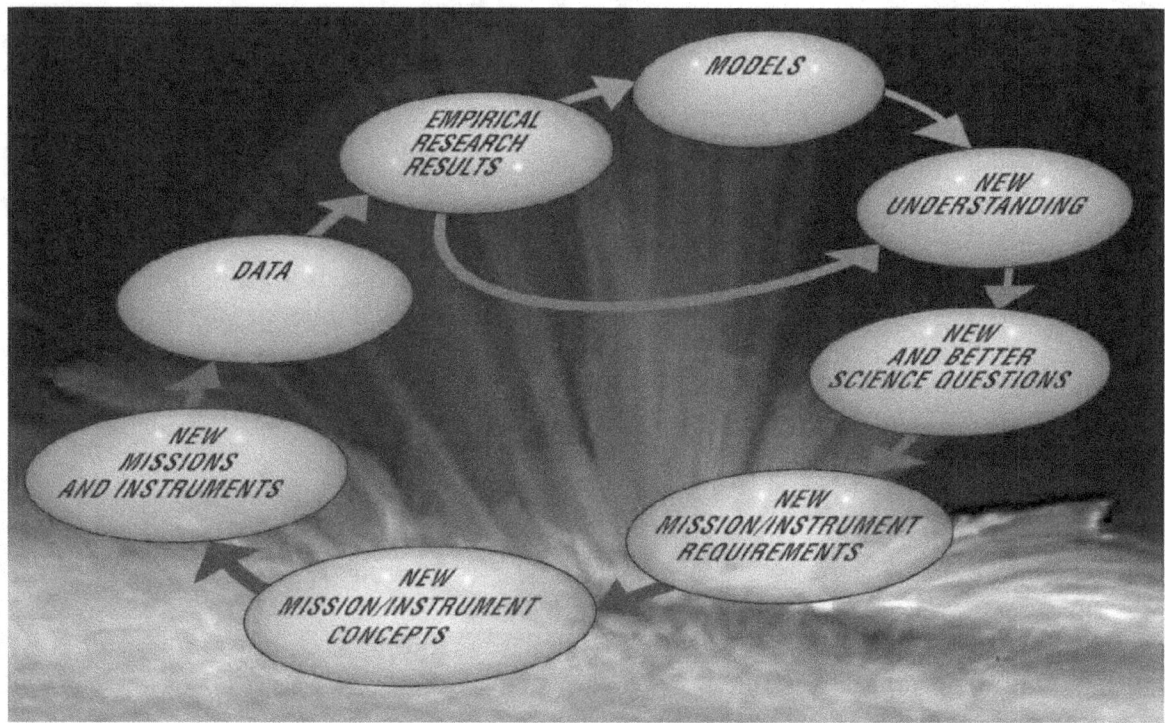

Today's special heliophysics science challenge is to consistently understand the physics and coupling of this "system of systems" over many spatial and time scales. Science information systems are key to drawing data and science both from individual science investigations and across the boundaries of the individual missions and instruments of the Heliophysics Great Observatory (HGO) now operated by NASA. In addition, NASA's Heliophysics Data Policy (see links on http://hpde.gsfc.nasa.gov) demands open and useful access to NASA data by the broad NASA and international science research community to ensure full science return to the public on NASA's investment in science missions and investigations.

The congruence of instrument design, physics, modeling, data analysis, and information technology expertise and experience in HSD makes it a natural center for leadership in the design and implementation of science information systems (SIS) working in close partnership with the external science and technology communities. Categories for systems currently defined in the Heliophysics Data Policy include the following:;

- o Mission/Investigation Data Facilities
- o Virtual Discipline Observatories (VxOs)
- o Resident and Final Archives
- o Science Data and Modeling Centers
- o Deep Archive(s)

The evolving heliophysics data vision is a distributed set of mission and investigation data facilities closely tied to the scientists responsible for active processing of data, coordinated with active archives and data centers joined together by Virtual Discipline Observatory (VxO) data location, and retrieval and user services to ensure easy consolidated access to usable data with views of the data customized to research discipline needs. In general, the HSD expects to leverage existing capabilities and services wherever practical, but to also support the introduction of new technology to enable a more distributed architecture. The first versions of the VxOs are intended to follow a "small-box" model focused on helping users find and retrieve fully described data relevant to their research needs (e.g., selected by times with parameters in a given data range from a selection of instruments).

Mission and investigation data facilities under HSD direction include

- o Solar Data Analysis Center (SDAC) support for SOHO and SOHO/EIT, solar instruments on STEREO, and data from TRACE, Hinode, and other missions
- o STEREO Science Center (SSC)
- o Critical support for SDO's distributed science archive and distribution system
- o Operation of the Wind-Geotail (Polar-Wind-Geotail) ground data system
- o Production of higher-level Wind (MFI, SWE electron, WAVES), Voyager (MAG CRS), and STEREO (SWAVES) instrument products
- o C/NOFS-CINDI data production

Two key HSD projects, the Space Physics Data Facility (SPDF) and SDAC serve as broad-based and heavily used, multi-mission, active archive and distribution projects for non-solar and solar imaging data respectively. The following graph shows SPDF service usage, where an execution is a request to create a customized plot, listing, or output file.

SPDF also maintains the Satellite Situation Center orbit and science planning service, as well as the Common Data Format (CDF) software, which is increasingly a standard, self-describing, self-documenting format for non-solar heliophysics data products. SPDF and SDAC (as noted above) also support multiple mission and investigation data distribution requirements as active science archives and serve as the final archives for older mission data.

HSD is a leading center of activity for Virtual Observatories, including the leadership of the newly formed Heliophysics Data and Modeling Consortium (HDMC) project for funding and coordination among the various VxOs and Resident Archives and direct leadership of the original set of VxOs selected by NASA:

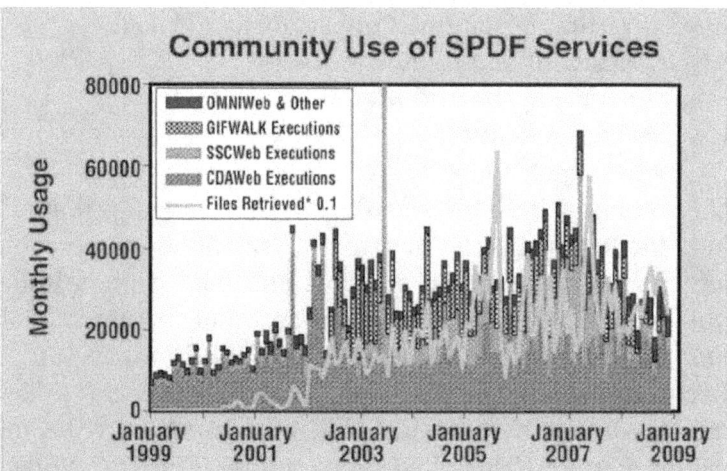

- o Virtual Solar Observatory (VSO)
- o Virtual Heliospheric Observatory (VHO)
- o Virtual Magnetospheric Observatory at Goddard (VMO/G)
- o Virtual Ionospheric/Thermospheric/Mesospheric Observatory (VITMO) (as Co-Is)

and more special-interest VxOs focused to special requirements in

- o Virtual Energetic Particles Observatory (VEPO)
- o Virtual Wave Observatory (VWO)
- o Heliospheric Event List Manager (HELM)

HSD staff also have active roles in defining the Space Physics Archive Search and Extract (SPASE) data model intended as a common language to allow queries among the VxOs. The SPDF project now supports the Virtual Space Physics Observatory interface, which is a need identified in the Heliophysics Data Policy to comprehensively identify and make "user findable" a full set of heliophysics data holdings across these disciplines leveraging both the present inventories of the individual VxOs and other information.

The overall goal is to make better data more readily findable and usable for researchers in a given discipline across the boundaries of specific missions, instruments, and times. This will be done by using an optimal mix of existing capabilities with appropriate standards and new technology to improve data access, and by using tighter coupling to the data-providing community to ensure data are fully and properly described for correct and independent future research use. HSD is looking forward to 1) closer coupling of data to models, 2) the more extensive use of distributed services through which data can piped for specific processing or other added value, 3) an ongoing effort to define lower-cost science ground-system design and implementation approaches, and 4) more sophisticated onboard data capabilities that can lead towards what are sometimes termed "sensorwebs" among active instruments.

COMMUNITY COORDINATED MODELING CENTER

The Community Coordinated Modeling Center (CCMC) is a US interagency activity aiming at research in support of the generation of advanced space weather models. The CCMC consortium consists of NASA, NSF[*], NOAA[†], the US Air Force (USAF) Weather Agency, Directorate of Weather, Space and Missiles System Center, the Air Force Research Laboratory, the Air Force Office of Scientific Research, and the Office of Naval Research. CCMC's central facility is located at GSFC. The CCMC is supported primarily by NASA and by NSF. At the present time, CCMC staff is 10 FTE strong, consisting of space and computer scientists and Information Technology professionals.

The first function of the CCMC is to provide a mechanism by which research models can be validated, tested, and improved for eventual use in space weather forecasting, such as needed for NASA's Vision for Space Exploration. Models that have completed their development and passed metrics-based evaluations and science-based validations are being handed off to the forecasting centers at NOAA and the US Air Force for space weather applications. In this function, CCMC acts as an unbiased evaluator that bridges the gap between space science research and space weather applications.

As a second, equally important function, the CCMC provides to space science researchers all over the world the use of space science models, even if they are not model owners themselves. This service to the research community is implemented through the execution of model "runs-on-request" for specific events of interest to space science researchers at no cost to the requestor. Model output is made available to the science customer by means of tailored analysis tools and data dissemination in standard formats. Through this activity and the concurrent development of advanced visualization tools, CCMC provides to the general science community unprecedented access to a large number of state-of-the-art research models. The continuously expanding model set includes models in all scientific domains from the solar corona to Earth's upper atmosphere.

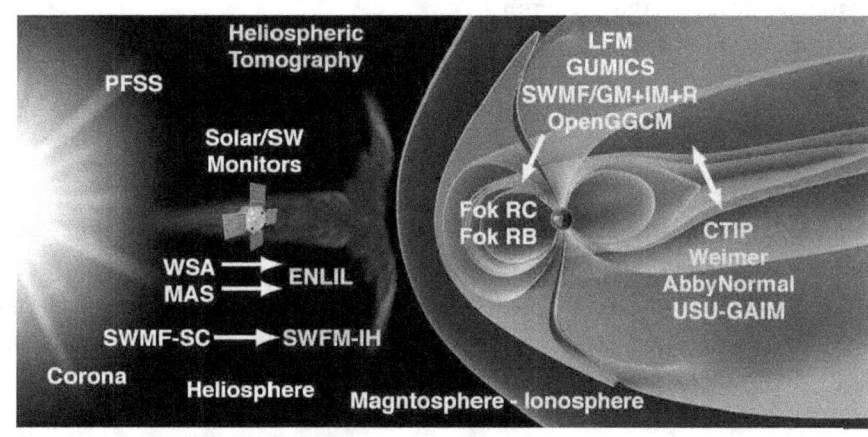

Overview of models at the CCMC.

---

[*] NSF: National Science Foundation
[†] NOAA: National Aeronautics and Space Administration

Science Support

CCMC science services are provided through Web access (http://ccmc.gsfc.nasa.gov). Here, users can request calculations from more than 25 modern space science models, which are hosted at the CCMC through very positive collaborations with their owners.

After a run request, a user will be notified via e-mail once the calculation is complete. At this time, the run can be analyzed via tailored visualization tools, again via Web access. These tools have been continuously refined for almost 10 years, primarily in response to user requests. Recently added capabilities include Poynting flux calculations, model outputs, field-line tracing along satellite trajectories, and movie generation.

The utilization of CCMC Run-on-Request services continues to grow rapidly.

CCMC run analysis services are also used heavily. In 2008, an average month saw 4,200 visitors, 19,200 visits, and 262,000 requests for pages. Visualization requests from 870 users led to the creation of 34,500 visualization pages each month. These monthly averages continue to increase from year to year.

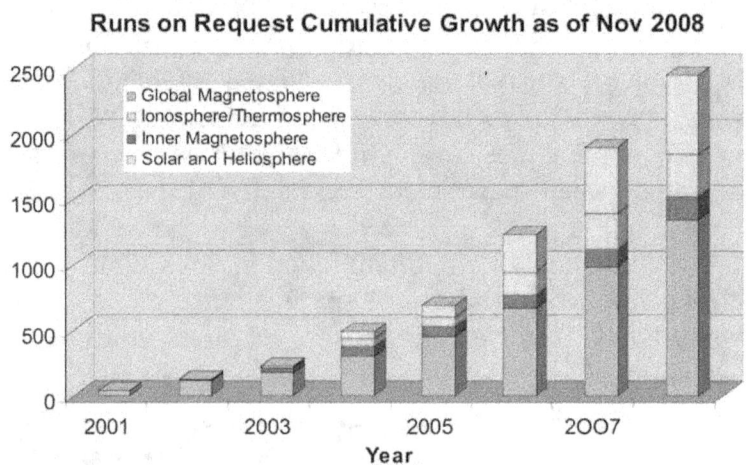

Cumulative RoR growth

Furthermore, the CCMC staff is continuing to develop new means to support users in their scientific studies. In the realm of visualization, OpenDx-based visualization with enhanced 3-D capabilities is now available for a number of different model outputs. With both OpenDx and IDL-based 3-D visualizations, users can request output in Virtual Reality Modeling Language (VRML) form, which permits real-time in-depth analysis of complex 3-D structures.

Further science services at the CCMC include the provision of the Model Web, where a large set of empirical or analytical models, such as the International Reference Ionosphere (IRI), are available for interested users for download or execution. CCMC is also supporting space science missions, for example STEREO and THEMIS. Science mission support includes background science information derived from routine runs, or calculations in support of specific campaigns or specific science objectives.

As in the past, the future of CCMC services will be shaped by the needs of the science community. Community input is solicited formally and informally, through meetings, tailored workshops, and personal contacts.

Space Weather Modeling

The second focus of CCMC is related to the need to bring modern space science modeling to bear on the needs of space weather forecasting. Forecasting agencies need models of proven forecasting abilities. Operational models must be robust, have accuracies that are well understood, and be packaged in a way that makes them easy to integrate into existing computational environments. The operational models must also be capable of executing reliably in real time. With these objectives, the CCMC tests models for accuracy, robustness, performance, and portability. Model accuracy is tested in two ways, through metric studies and through scientific validation studies. As a result, the CCMC can characterize the usefulness of these models for the forecasting role to which they would be applied. By testing how the models behave with a wide range of inputs, the CCMC can establish regimes in which the models appear robust and can identify regimes that cause the models to fail. Through exposure to a range of compilers and hardware platforms, the CCMC improves the portability of these models. Finally, by measuring computational performance, the CCMC establishes the conditions needed for each model simulation to complete in real time or better.

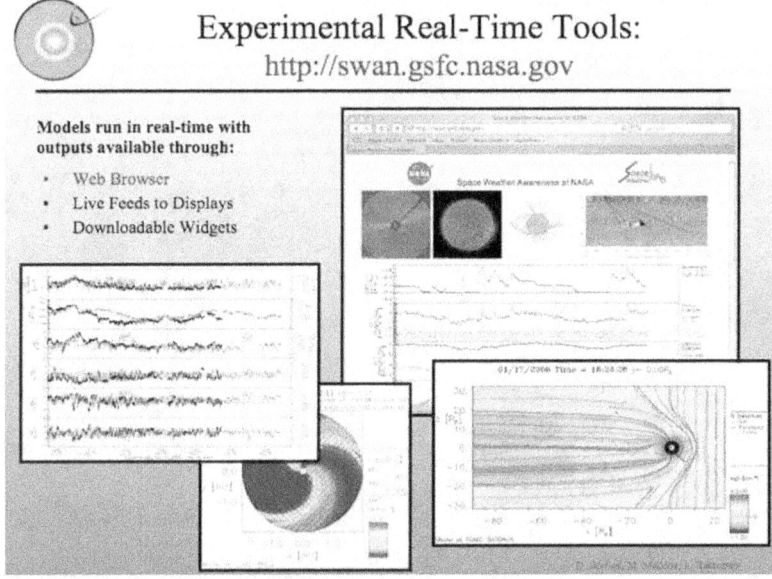

Sample plots from SWAN site.

Performing these types of analyses on models in a protected academic environment might produce less than objective results. It is imperative that these models undergo tests in an environment that stresses them in the same fashion as will occur at the operational agencies. Accordingly, the CCMC challenges all models by utilizing them in a quasi-operational setting, namely the Runs-on-Request facility, as well as through a real-time simulation system. For the latter purpose, CCMC has established various real-time execution systems, starting with solar and solar wind modeling of photospheric magnetograms, and magnetospheric calculations driven by ACE data. The establishment of these systems has led to a wealth of expertise in the design and maintenance of real-time modeling. In collaboration with model owners, CCMC has created robust models of scientific and space weather utility.

As a byproduct of real-time model execution, the CCMC staff has developed various model products, or "tools," which provide space weather-relevant information to interested parties. Driven by a mandate from NASA/HQ, CCMC has begun to provide and collect data sources of relevance to the NASA Exploration Initiative, and to NASA

Space Science Mission Operations. This information is provided to any interested entity on the Space Weather Awareness at NASA (SWAN) Web site: http://swan.gsfc.nasa.gov.

Owing to the diversity of space weather-relevant information sources, space weather analysis and forecasting will, in the future, increasingly rely on access to distributed information providers. The SWAN site is a first step in this direction of information collection. While it, along with further development activities supported by NASA's Office of Chief Engineer, addresses NASA's needs primarily, this information can also support governmental as well as commercial space weather interests. CCMC, therefore, has ongoing collaborations with USAF, NOAA, international space weather interests, and commercial enterprises in the US. An example of the latter is the Electrical Power Research Institute (EPRI), which, in a jointly funded activity, collaborates in the evaluation of space weather modeling products for power grid uses.

More information about the CCMC may be found at the CCMC Web site http://ccmc.gsfc.nasa.gov.

TECHNOLOGY DEVELOPMENT

Goddard Space Flight Center and NASA in general support new technology in the Heliophysics division through three programs. The main source of support is the GSFC Internal Research and Development (IRAD) program, which funds the development of new technology to support proposals to be submitted to future announcements of opportunity. The Technical Equipment program provides funding for purchases of advanced technology equipment and replacement of outdated equipment needed to support the development of new instrumentation. The Small Business Innovative Research (SBIR) program provides funding to develop new flight hardware and software concepts in the commercial sector in support of future flight opportunities.

The IRAD program of GSFC runs on an annual basis for each fiscal year. An announcement of opportunity is released in early June describing the requirements of the opportunity for the subsequent fiscal year. Step-1 proposals are submitted in late June. A proposal review panel lead by Dr. John Sigwarth, the Chief Science Technologist for the Heliophysics Science Division, meets in July to rank the proposals and recommend to the Center-wide integration panel those proposals that should progress to the second phase. Step-2 proposals are submitted in late July. The review panel meets again in early August to rank the Step-2 proposals and recommend funding to the Center-wide integration panel. Funding approval by the Center-wide integration panel is contingent on the available resources and the relative ranking of the Heliophysics Science Division proposals with respect to other focus areas within GSFC. In FY08, a total of 17 Step-1 proposals were submitted, resulting in 7 projects funded by the GSFC IRAD program. An additional project was added halfway through the year. For these winning projects, the FY08 IRAD funded a total of 11.7 FTEs and an additional $606.5 K for procurements in support of the selected IRAD proposals. In FY09, a total of 15 Step-1 proposals were submitted resulting in 7 projects funded by the GSFC IRAD program. For these winning projects, the FY09 IRAD program funded a total of 11.8 FTEs and an additional $434.5K for procurements in support of the selected IRAD proposals.

The new technologies supported by the GSFC IRAD program cover a broad range of heliophysics topics. These technologies include new neutral-to-ion conversion surfaces in a venetian blind configuration for miniaturization of neutral-atom imagers; plasma impedance spectrum analyzers with state-of-the-art electronics to process the observations onboard and determine the ionospheric plasma density and temperature at high cadence; a Far Ultraviolet (FUV) Fabry Perot interferometer, extremely high line density reflective grating, and electron-multiplying Charged Coupled Devices (CCDs) for measuring the Doppler-broadened line width of thermospheric emissions; the demonstration of the next-generation coronagraph that will extract additional information about electron temperatures and flow speeds from observations of the visible light spectrum; design of a new low-mass, low-power ionospheric sounder; and development of a data compression application-specific integrated circuit (ASIC), fast-stepping high voltage power supplies, and particle detector optics.

Three of the instruments selected for FY08 IRAD funding have involved students of the US Naval Academy. Because of this cooperation, the Thermospheric Temperature Imager (TTI), the Plasma Impedance Spectrum Analyzer (PISA), and the Miniature Imager for Neutral Ionospheric atoms and Magnetospheric Electrons (MINI-ME) have been briefed to the Department of the Navy and the Department of Defense (DoD) Space Experiment Review Boards (SERB) and selected to be included on the respective SERB lists. As a direct consequence of this inclusion on the SERB list, TTI, PISA, and MINI-ME will fly on a Space Test Program spacecraft in December 2009.

The GSFC Technical Equipment program is run on an annual fiscal year basis to purchase new capital equipment or replace obsolete equipment. Each year suggestions for equipment purchases are solicited from HSD members. These requests are ranked by the HSD management and brought forward to the Science and Exploration Directorate. In FY08, approximately $200K was allocated for technical equipment purchases within HSD. In FY08, the technical equipment fund within the HSD was oversubscribed by 300%.

The SBIR program within NASA provides an opportunity for small, high technology companies to participate in Government-sponsored research and development (R&D) efforts in key technology areas. New technology needs for HSD are identified by members within the division and vetted by GSFC- and NASA-wide integration panels. Solicitations based on the identified key technology areas are released annually. Selected SBIR companies work with the oversight of the Contracting Officer's Technical Representative to achieve the desired technology goals. For FY08, two SBIR proposals were selected in the area of heliophysics.

APPENDIX 1: INDIVIDUAL SCIENTIFIC RESEARCH

The following section contains some short summaries of the research work of the scientists working in HSD. The organization is divided up into four groups, or laboratories, with the civil servant complement shown in the detailed organization chart shown below.

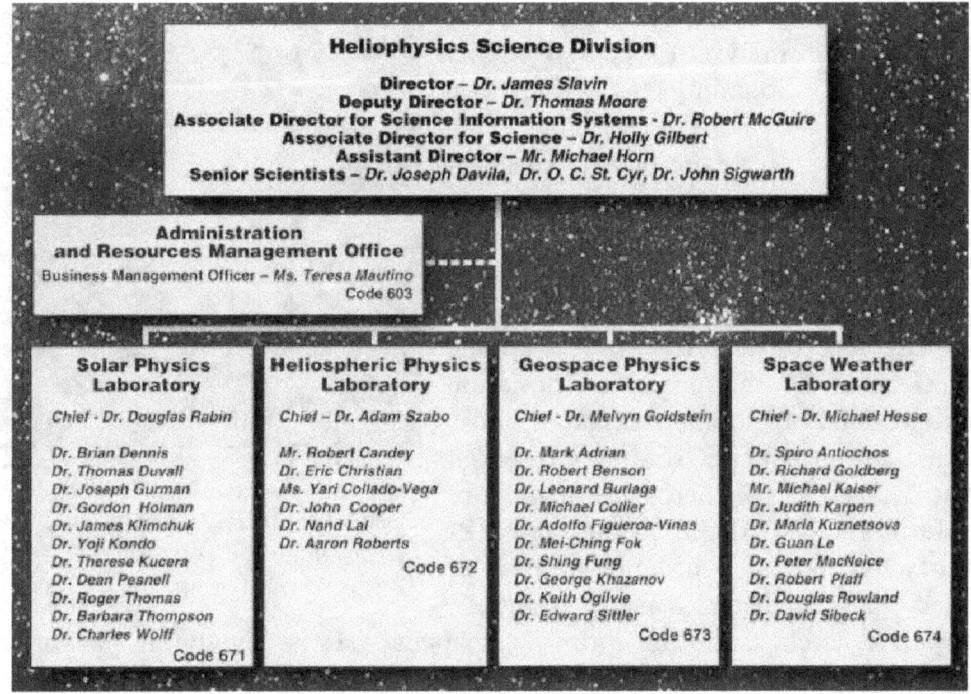

Mark A. Adrian

A thermospheric-ionospheric-magnetospheric physicist who joined Code 673 in November 2004, Dr. Adrian is the Deputy Project Scientist for the MMS satellite constellation mission and is the Instrument Scientist for the Dual Electron Spectrometer (DES) component of the MMS Fast Plasma Instrument (FPI). Dr. Adrian's research interests are focused primarily on the energization and acceleration of thermal to low-energy electrons throughout the heliosphere. He continues analysis of data from the Sounding of the Cleft Ion Fountain Energization Region (SCIFER) sounding rocket, probing the physics of high-latitude particle acceleration. In particular, Dr. Adrian continues analysis of data obtained using the Thermal Electron Capped Hemisphere Spectrometer (TECHS) flown on SCIFER in order to quantify the role of the thermal electron distribution in the formation of ion conics and outflows in the cusp. He is conducting plasmaspheric dynamics research using the Imager for Magnetopause-to-Aurora Global Exploration (IMAGE) EUV and RPI to quantify the development and evolution of embedded plasmaspheric density channels, plasmaspheric notches, and the entire plasmapause. He is developing plasma diagnostic hardware through analysis of the TECHS design concept, as well as developing plasma diagnostics for the, solar wind composition, neutral atmospheres and the plasmas associated with the development of sprites above convective thunderstorms.

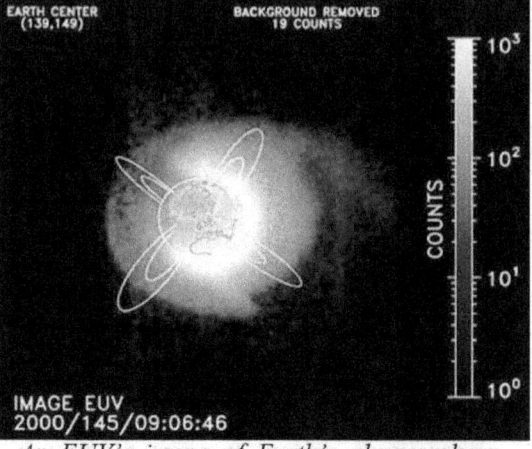

An EUV's image of Earth's plasmasphere displays a shoulder in the pre-noon plasmapause and plasmaspheric drainage plume in the dusk sector.

Spiro Antiochos

Dr. Antiochos has recently been working on extending the breakout model for CME initiation to fully 3-D topologies, development of a model for coronal jets, derivation of theorems on the topology, and extension of the thermal non-equilibrium model for prominence formation to non-steady heating. He has given invited presentations at the Fall 2007 AGU, Space Policy Institute Workshop, 2nd Heliospheric Network Workshop, Spring '08 SPD, and the Hinode-2 Science Meeting. He supervises the Ph.D. thesis research of a graduate student from the University of Michigan (where

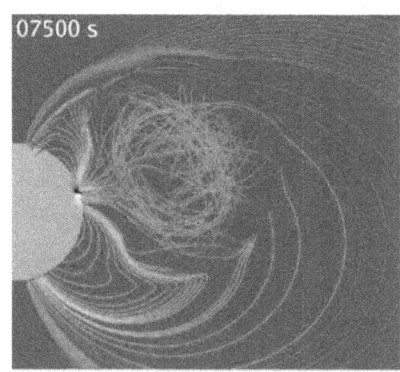

3-D breakout simulation from Lynch et al., 2008

Dr. Antiochos is an adjunct professor) and presented two lectures at the 2008 Summer School in Solar Physics. He was elected a fellow of the American Physical Society.

Robert Benson

Dr. Benson continued his research in plasma-wave emissions, wave-particle interactions and wave propagation in the ionosphere and magnetosphere. Three publications (one in print, two submitted) and three presentations resulted from this work in the last year. These were based on his former role as a participating scientist on the Radio Plasma Imager (RPI) on the IMAGE satellite (IMAGE is no longer active), his role as PI on a Heliophysics Living With a Star Targeted Research and Technology (LWS TR&T) award concerning the investigation of large mid-latitude topside-ionospheric/plasmaspheric gradients, and his role as PI on a project to preserve and transform a representative portion of the International Satellites for Ionospheric Studies (ISIS) topside-sounder data from an analog to a digital format. The RPI work reviewed some fundamental space plasma physics accomplishments that were based on space-borne radio sounding. The LWS research revealed the altitude structure of mid-latitude ionospheric features associated with magnetic-field lines through the plasmapause boundary. The ISIS data restoration effort has brought the number of digital ionospheric topside ionograms from the time period between 1965 and 1985 to more than ½ million; essentially creating a new satellite mission with old data. The distribution of these data records from the Alouette-2, ISIS-1, and ISIS-2 satellites over an interval of nearly two solar cycles (data available from the National Space Science Data Center [NSSDC]), and is illustrated in the lower panel of the figure; the upper panel shows the distribution of topside electron density profiles (also available from the NSSDC), many based on the new digital ionograms (those labeled TOPIST[*]).

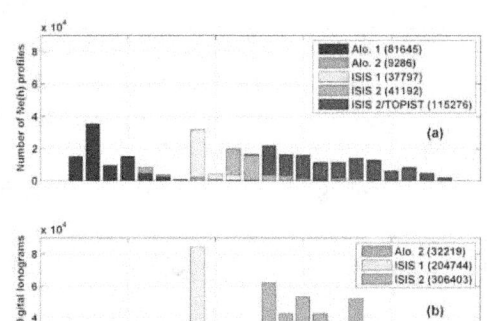

Anand Bhatia

Dr. Bhatia gave the "Professor S.C. Sircar Memorial Lecture" at the Indian Institute for Cultivation for Science in Kolkta (Calcutta), India on 2008 April 11. The title of the talk was "Polarizabilities, Rydberg states, and scattering of electrons from hydrogenic systems." He was awarded an engraved silver plate to commemorate the occasion. In addition, he gave a talk on "Hybrid theory for the scattering of electrons from hydrogenic systems," in Tallahassee, Florida on 2008 June 14.

---

[*] TOPIST: Topside Ionospheric Scaler with True height (algorithm)

Dieter Bilitza

A research professor at George Mason University (GMU) working in Goddard's Heliospheric Physics Laboratory, Dr. Bilitza is an expert in ionospheric physics. He is the principal author of the International Reference Ionosphere (IRI), a model widely used in the science and engineering community, which recently also became an International Standards Organization (ISO) Technical Standard. IRI is used in a number of NASA technical documents to define the ionospheric environment and is the model recommended by the European Cooperation for Space Standardization (ECSS). In his most recent efforts, he has used a large database of satellite *in situ* measurements, the majority from NASA satellites, to study the solar activity variations of densities and temperatures in the topside ionosphere with the goal of improving the representation of these variations in IRI. Professor Bilitza was awarded a NASA Space Science Achievement Award in 2007.

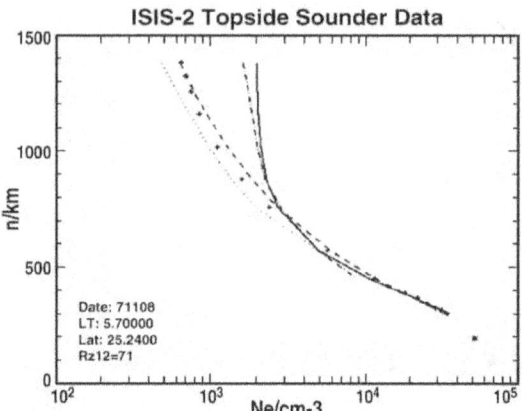

The figure compares measurements and IRI model predictions for the topside electron density. Significant discrepancies are seen above 800 km altitude with the IRI-2001 model (solid line) overestimating the data (+) obtained with the ISIS-2 topside sounder satellite. A correction term introduced in IRI-2007 (dotted line) produces much better agreement.

Scott Boardsen

Dr. Boardsen, of the University of Maryland, Baltimore County (UMBC), analyzed MESSENGER[*] magnetometer data from Mercury flybys 1 and 2, specifically studying narrow band ultra-low frequency (ULF) waves that were detected above helium + cyclotron frequency inside Mercury's magnetosphere. He worked with Dr. Nick Omidi at Solana Science, Inc., on hybrid simulations of Mercury that includes sodium+ created by photo-ionization in the sodium exosphere.

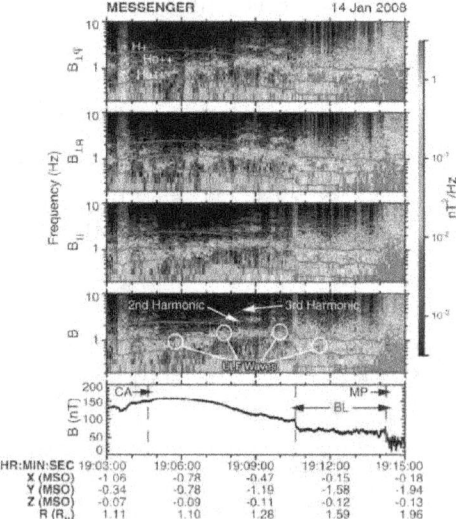

Dynamic Spectra of ULF waves and their harmonics, observed outbound from closest approach, detected during the first MESSENGER Mercury flyby. The white lines indicate the He+, He++, and proton cyclotron frequencies.

---

[*] MESSENGER: Mercury Surface, Space Environment, Geochemistry and Ranging (NASA's mission to Mercury)

Ben Breech

Dr. Breech recently joined Goddard in September 2008 as an NPP fellow after completing Ph.D. degrees in Computer Science and Physics at the University of Delaware. His research focuses on modeling turbulence transport within the heliosphere. He is currently working on modeling effects of electron heat conduction on turbulence and extending phenomenological turbulent transport models to the solar corona. He has presented results from his work at AGU meetings.

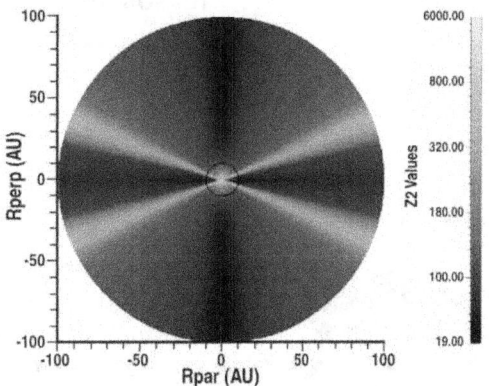

Distribution of turbulent energy in the heliosphere.

Kenneth Bromund

Mr. Bromund is a software engineer with SP Systems, Inc. He has been working at GSFC with Dr. Rob Pfaff on *in situ* measurements of Electric and Magnetic Fields using the Vector Electric Fields Instrument (VEFI) on the Communications/Navigation Outage Forecast System (C/NOFS) satellite. He investigated the effects of the spacecraft environment on the VEFI measurements, and developed algorithms to identify data that has been corrupted. He developed and delivered to the Air Force Research Laboratory (AFRL) software that will be used to produce near-real time electric and magnetic field solutions from the VEFI data. AFRL will incorporate the software Mr. Bromund developed into a system designed to now-cast and forecast scintillations in the equatorial ionosphere.

Jeff W. Brosius

Drs. Brosius and Holman investigated the thermal and dynamic evolution of a compact solar microflare over a wide temperature range at rapid cadence. The microflare's properties and behavior are those of a miniature flare undergoing gentle chromospheric evaporation, likely produced by reconnection-driven nonthermal electrons. Drs. Brosius, Rabin, Thomas, and Landi derived a coronal bright point's DEM from a well-calibrated EUV spectrum obtained with EUNIS-06. Photospheric (not coronal) element abundances were required to achieve equality and consistency in the transition region DEM derived from lines with a low first ionization potential (FIP) and lines with a high FIP. Drs. Brosius, Rabin, and Thomas investigated a transient brightening in the quiet Sun observed at rapid cadence (2.1 s) with EUNIS-06. The transient's measured properties are consistent with its identification as a blinker or an elementary blinker, and its observed behavior suggests a formation mechanism involving gentle chromospheric evaporation. Jess et al. (2008) investigated transition region velocity oscillations that correspond to MHD fast-body global sausage modes in an active region loop arcade observed with EUNIS-06. Keenan et al. (2008) compared theoretical atomic physics calculations for Fe X with sounding rocket EUV spectra.

The EUNIS sounding rocket team received the Robert H. Goddard Exceptional Achievement Award for Science.

Matthew Burger

Dr. Burger is an associate research scientist with the Goddard Earth Sciences and Technology (GEST) Center at the University of Maryland, Baltimore County. He is working with the Cassini Plasma Spectrometer team at GSFC to measure the plasma in Saturn's inner magnetosphere and to understand the interactions between Saturn's magnetosphere and icy satellites. In particular he has worked to understand the interaction between plasma in the magnetosphere and the water plume at Enceladus' south pole. He has also made ground-based observations showing a surprising lack of sodium in the plume places strong constraints on the source mechanism for the plume.

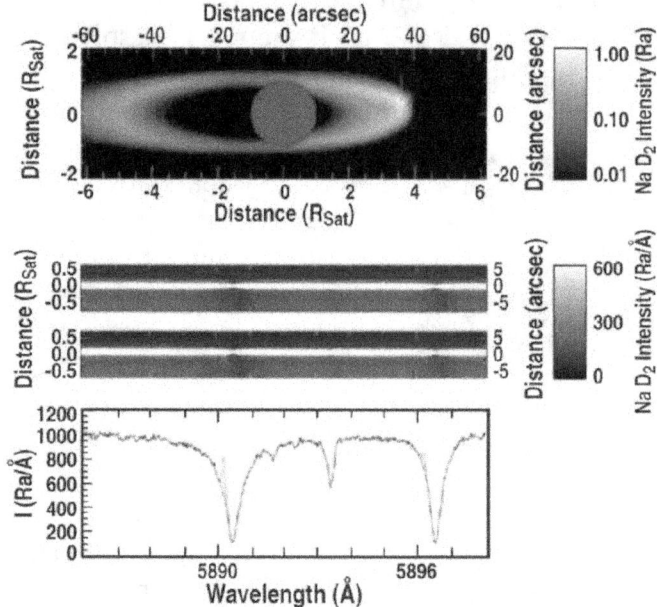

Model of an Enceladus sodium plume assuming a sodium escape rate of 0.4 g/s, corresponding to a Na/H$_2$O mixing ratio of $2x10^{-6}$. The green line represents the slit across Enceladus. Sum of all sky-subtracted spectral images for Enceladus observations with (bottom) and without (top) simulated sodium plume emission. The simulated signal does not include photon noise. Spectrum of the spatial region indicated by the arrow in (b). The black curve is the data alone; the red curve adds 40 R of simulated emission derived from the model in (a). This simulated emission would be readily detectable. The actual upper limit from these data is 8 R, yielding an Na/H$_2$O$<4x10^{-7}$.

Leonard Burlaga

Dr. Burlaga is an astrophysicist specializing in the structure and dynamics of the heliosphere and its interaction with the interstellar medium. He is a Co-I on the magnetic field and plasma instruments on Voyager 1 and 2, the SECCHI instrument on STEREO, the magnetic field instrument on ACE, and the plasma instrument on Wind. In addition, he is on a theoretical team studying the dynamics of the distant heliosphere. Perhaps the principal result this year is the discovery of the termination shock by Voyager 2. In a paper published in *Nature* this year, Dr. Burlaga and his co-authors demonstrate that the termination shock is a dynamic feature, reforming on a scale of a few hours, with the basic features of a quasi-perpendicular shock modified by pickup protons. Dr. Burlaga also presented an invited paper at the ASTRONUM Conference summarizing his work demonstrating the fundamental importance of Nonextensive Statistical Mechanics in the dynamics of the heliosphere. The National Academy of Sciences awarded Dr. Burlaga the Arctowski Medal "for pioneering studies of the magnetized solar wind plasma from 0.3 to 102 AU, including the recent crossings of the Voyagers of the heliospheric termination shock and their entry in the heliosheath."

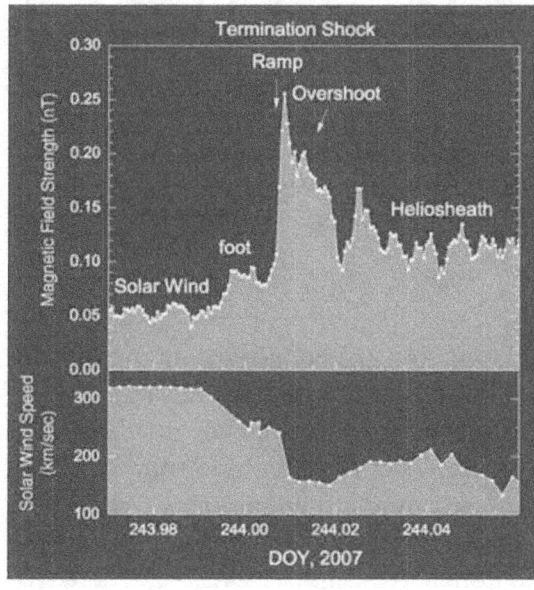
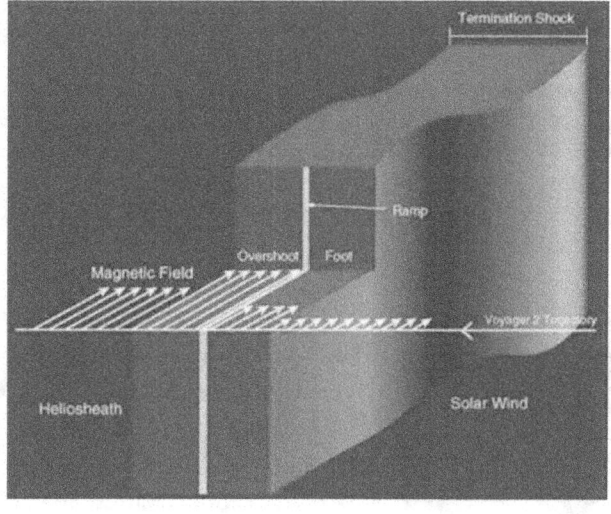

Profiles of the magnetic field strength and solar wind speed showing the structure of the termination shock.

Natalia Buzulukova

Dr. Buzulukova joined Code 670 in May 2007 as an NPP fellow. She is doing research on inner magnetosphere modeling and data analysis. Her scientific advisers are Dr. T.E. Moore and Dr. M.-C. Fok. During 2007–2008, she has had extensive experience in running ring current/plasmasphere model (CRCM model, code developer M.-C. Fok). Using modeled results and EUV IMAGE data of the plasmasphere, she did a detailed analysis of plasmaspheric undulations was done (17 April 2002 event). She also has modified the original CRCM code to couple with the MHD code (BATSRUS model). This project is conducted under the partnership with the CCMC at Goddard. She is also involved in the TWINS project activity. During the COSPAR 37th Assembly in Montreal, Canada, July 2008, she was the Deputy Organizer of the section titled "Ionosphere-magnetosphere coupling: the role of Alfvén waves in auroral processes."

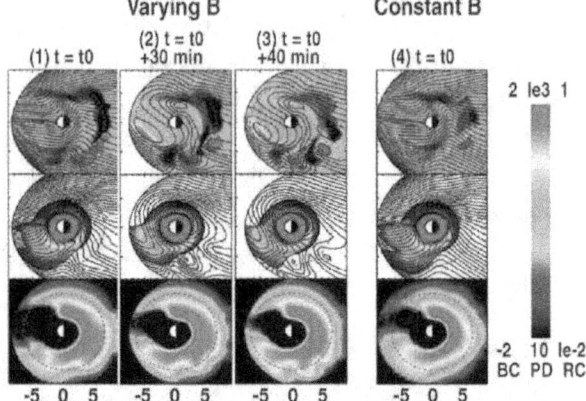

Formation of plasmaspheric undulations. Columns 1–3 correspond to three times during the substorms on 17 April 2002. The rows show simulated Region 2 Birkleland current (BC) mapped to the equator (top panels), plasmasphere density (PD) (middle panels) and ring current (RC) pressure for 16–27 keV H+ (bottom panels). Column 4 is the simulation result with constant magnetic field configuration. The dashed cycles are geosynchronous orbits. Potential contours without and with corotation are overlaid in the top and middle panels, respectively.

Mike Calabrese

Mike provides integration, coordination, scheduling, analysis, and assessment in support of the Heliophysics advanced mission studies and technology assessments for HSD. He attended an all-day HSD Meeting on Future Heliophysics Science Flight Opportunities 2008 April 7, and provided assessment and recommendations to HSD. He provided a same-day response to a NASA HQ request for electronic versions of 12 Heliophysics Roadmap Mission Studies conducted in 2005 at GSFC and the Jet Propulsion Laboratory (JPL). He attended the NASA Heliophysics Town Hall Meeting in College Park, Maryland on 2008 May 19–20, to provide background for Heliophysics Advanced Mission Planning. Approximately 100 community attendees provided input, which included over 30 Roadmap Mission posters for the 2009 Heliophysics Strategic Planning Roadmap Team, which placed emphasis on a sustainable and flexible plan to be delivered 2008 November 1. He participated in breakout sessions to identify science objectives and technology needs, and he attended the National Academy of Sciences Colloquium on "Forging the Future of Space Science for the Next 50 Years" on 2008 June 26 to provide context on advanced Space Science Missions identification.

Robert Candey

Robert Candey is the Chief Architect for the Space Physics Data Facility, manages the science ground system for the Wind and Geotail (and previously Polar) spacecraft, and leads the new Heliophysics Event List Manager (HELM) VxO project and the existing xSonify sonification and Common Data Format (CDF) projects. He supports the VxOs, particularly VITMO and VWO. He led the very successful Dynamics Explorer data restoration effort with Dieter Bilitza and Howard Leckner. He gave presentations at the Fall AGU meeting, mentored Wanda Diaz, participated with the MMS E/PO team, supported the Combined Federal Campaign (CFC) as the Code 672 keyworker, and was on the 670 Peer Award committee. He helped to host a visit with Iku Shinohara and Ken Murata from Japan's Institute for Space and Aeronautical Science (ISAS) and Takaaki Matsuzawa from the Washington office of the Japan Aerospace Exploration Agency (JAXA).

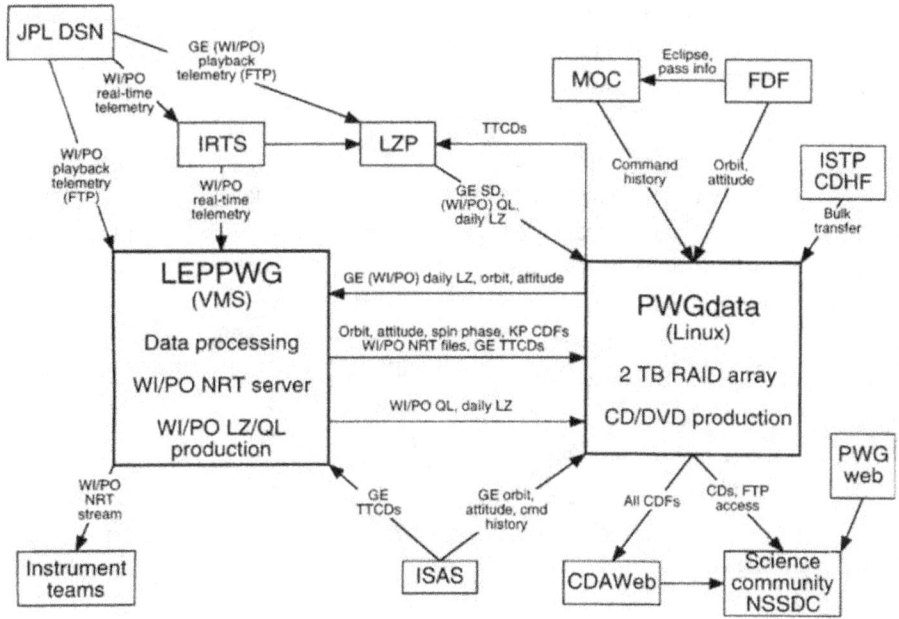

The Polar-Wind-Geotail (PWG) science ground system automatically handles input from many data providers using a simple directory-based work flow system based on small Perl scripts.

Peter Chen

P.C. Chen, in collaboration with D. Rabin (Code 671), continued work to develop lightweight mirrors made of polymer composite materials and to study their long term stability. A process has been developed to enable the fabrication of telescope-quality mirrors using specially formulated epoxy resins, carbon fibers, and carbon nanotubes. A number of mirrors with diameters of 0.1–0.2 m have been fabricated having excellent optical figures as measured using a medium resolution (100 lines/inch) Ronchi grating. An interferometer is being constructed. Parts are being fabricated to make composite mirror telescopes for field testing. Chen, in collaboration with D. Rabin, and M.Van Steenberg (Code 604), developed a method to make potentially very large telescopes on the Moon using a combination of lunar dust stimulant JSC-1A, carbon nanotubes, and epoxy resins. A 0.3 m mirror was successfully fabricated. The unit was exhibited at the AAS meeting in St. Louis, Missouri, and was the subject of a NASA press release, as well as an article in science.nasa.gov.

Sheng-Hsien (Sean) Chen

Dr. Chen is a research scientist with the Universities Space Research Association (USRA) working in Geospace Physics Laboratory. Dr. Chen is analyzing Polar, Cluster, and THEMIS plasma and magnetic field data to study the interaction of solar wind, magnetospheric, and ionospheric plasmas at the magnetospheric boundaries. More recently, Dr. Chen has been working on the long-term solar cycle effect of the solar and geomagnetic activity on the ionospheric outflows in the polar cap regions.

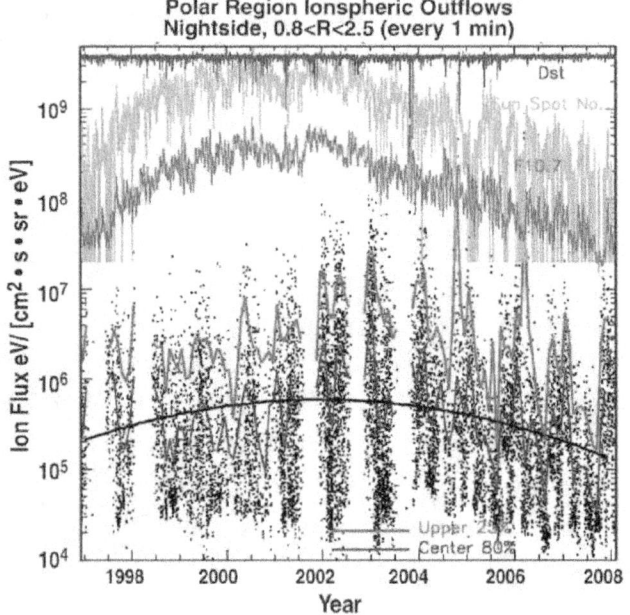

Variations of ionospheric outflows at the top of the ionosphere in the polar cap regions over the 11-year solar cycle between 1997 and 2007.

Eric Christian

Eric Christian arrived at Goddard Space Flight Center in November after a six-year stint as a Program Scientist at NASA Headquarters. He will be studying energetic particles coming from the Sun, the heliosphere, and the galaxy with a wide range of missions. He is the Deputy Project Scientist for the Solar-Terrestrial Relations Observatory (STEREO) and the Advanced Composition Explorer (ACE), and the Deputy Mission Scientist for IBEX. He is a member of the science team for the STEREO/IMPACT[*], ACE/CRIS[†], and ACE/SIS[‡] instruments and will be involved with data analysis from them with a primary goal of understanding the acceleration of solar energetic particles. He will also be working with the Voyager/CRS[§] and IBEX teams towards the scientific goal of understanding the origin of anomalous cosmic rays and the structure of the interaction of the heliosphere with the local interstellar medium.

Eric is a Co-I on the Trans-Iron Galactic Element Recorder (super-TIGER)—a newly selected balloon-borne instrument that will study ultra-heavy galactic cosmic rays, with a first Antarctic flight in 2012. He will help design, develop, and integrate the detectors and hardware that are Goddard's contribution to TIGER.

Eric has been active in NASA's Education and Public Outreach (E/PO) program for his entire career and will continue this in his new position. He is a Co-I on a recently selected Education and Public Outreach for Earth and Space Science (EPOESS) grant that will develop a particle and plasma multi-media library for the Web, and will be a participant on the upcoming Sun–Earth Day Web cast, as well as other E/PO projects.

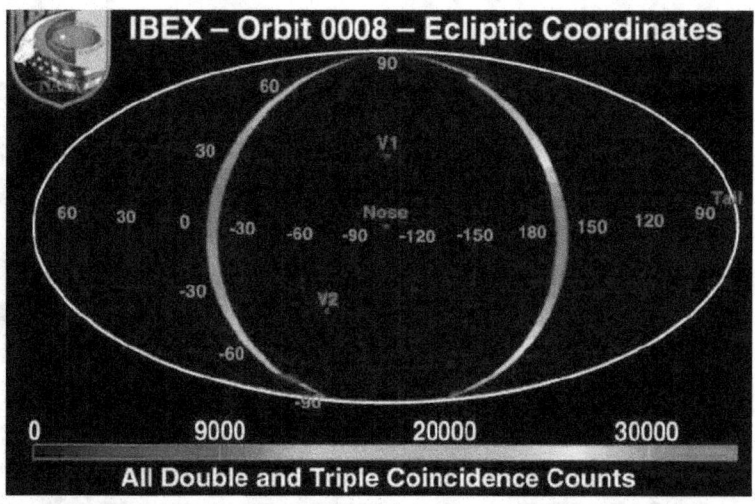

"First light" from the IBEX mission, showing the observed intensities of Energetic Neutral Atoms (ENAs) of around 1 keV in one slice of the sky. Backgrounds have not be subtracted and the high intensity region on the right is predominantly due to ENAs from Earth's Magnetosphere.

---

[*] IMPACT: *In situ* Measurements of Particles and CME Transients
[†] CRIS: Cosmic Ray Isotope Spectrometer
[‡] SIS: Solar Isotope Spectrometer
[§] CRS: Cosmic Ray Subsystem

Michael R. Collier

Dr. Collier has contributed to fabricating, calibrating, commanding, and analyzing data from the Low Energy Neutral Atom (LENA) imager on the IMAGE spacecraft. He studies primarily low energy neutral atoms in Earth's vicinity and participates heavily in the development of high time resolution particle instrumentation. He has analyzed Voyager 1 and 2 energetic particle data both in the interplanetary medium and in the magnetospheres of the outer planets. In addition, he studied Voyager 1 and 2 magnetometer data in, and in the proximity of, Jupiter's magnetosphere. He was also involved with hardware projects, performing numerical simulations to determine the design characteristics and dimensions of a solar wind composition instrument later launched on the Wind spacecraft, and testing instrument prototypes. He analyzed data from the three Wind sensors, particularly the MASS instrument, to determine abundance ratios, temperatures, and distribution functions of minor ion species in the solar wind. He has analyzed IMP 8, Wind, and Geotail magnetometer and particle data to study the effects of pressure discontinuities on the terrestrial magnetotail, to characterize interplanetary magnetic field correlations and absolute agreement, and to examine the properties of magnetic clouds, particularly shocks internal to magnetic clouds. He serves as Deputy Project Scientist for the Wind Spacecraft and as Instrument Scientist for IMAGE/LENA.

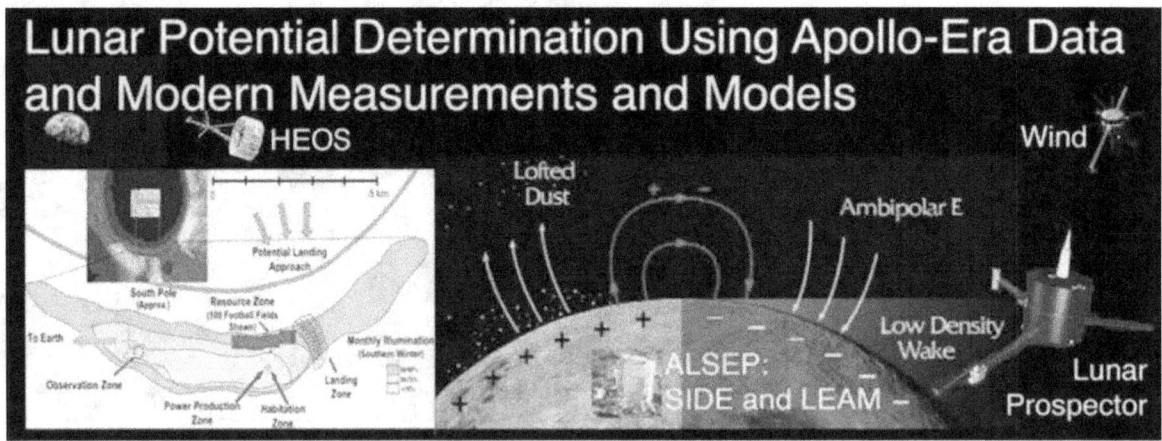

John Cooper

Dr. Cooper is the Chief Scientist for the Space Physics Data Facility (SPDF) within the Heliospheric Physics Laboratory. He advises SPDF on heliophysics science requirements for ongoing and future development of SPDF data systems, represents SPDF at science team meetings for various operational missions, notably IBEX now launched and operational in 2008, and works with SPDF staff on connecting SPDF to the emerging heliophysics data environment of NASA-supported virtual observatories. Dr. Cooper is the PI for the Virtual Energetic Particle Observatory (VEPO) serving the heliophysics data user community as a focus group component operating within the domain of the Virtual Heliospheric Observatory (VHO) for improved access and usability of energetic particle data products. He has pursued personal research interests on space radiation environment interactions with icy moons and Kuiper Belt Objects. He is also a member of the Cassini Plasma Spectrometer team at Goddard. Related to the Outer Planets Research (OPR), Dr. Cooper collaborated on the successful proposal of the University of Hawaii at Manoa to the W.M. Keck Foundation for the W.M. Keck Research Center in Astrochemistry. He is also a U.S. collaborator on the European Space Agency (ESA) Cosmic Visions proposals for the Laplace mission to Jupiter and the Tandem mission to Saturn, also contributing to the final report for the NASA flagship mission study on the Jupiter System Observer. He also participated in the International Heliophysical Year (IHY) and International Polar Year (IPY) 2007–2009 activities as a member of the ICESTAR[*]/IHY team and was chairperson and lead organizer for the Polar Gateways Arctic Circle Sunrise 2008 science-education conference at Barrow, Alaska on 2008 January 23–29. Polar Gateways was notable as a "green" conference with more than half of the science contributors participating remotely via video or phone conference from other sites in Sweden, Norway, Russia, and from universities and NASA Centers in the United States.

Web site banner, Polar Gateways Arctic Circle Sunrise 2008 Conference, Barrow, Alaska, 2008. New conference center (right), sea ice, sunrise, aurora, snow fox and owl, and whalebone objects are from photography at the conference. The old building (left) image is from the original IPY expedition to Barrow in 1881. (From the conference Web site at polargateways2008.gsfc.nasa.gov.)

---

[*] Interhemispheric Conjugacy Effects in Solar Terrestrial and Aeronomy Research

Joseph Michael Davila

Dr. Davila is currently a senior scientist in the Heliophysics Division at Goddard Space Flight Center in Greenbelt, Maryland. His research interests include the linear and non-linear theory of hydromagnetic waves; hydromagnetic instabilities due to energetic particle beams, resonance absorption in inhomogeneous plasmas, the acceleration of high speed wind streams in solar and stellar coronal holes, and plasma heating in closed magnetic structures. He has also published research on the acceleration of cosmic rays, the transport of energetic, particles within the galaxy, the modulation of galactic cosmic rays by the solar wind and the propagation of solar cosmic rays in the interplanetary medium.

Dr. Davila was the PI for the Solar Extreme-ultraviolet Research Telescope and Spectrograph (SERTS) from 1992 to 2002. He was Study Scientist for the STEREO Science Definition Team, and the Project Scientist for STEREO until 2004. Dr. Davila is the lead scientist for COR1 on the STEREO mission, and was responsible for building the instrument. He currently leads the COR1 data analysis team. He is a Co-Investigator and participated in the development of the Hinode Extreme-ultraviolet Imaging Spectrometer (EIS). He is currently engaged in data analysis from EIS.

Dr. Davila has led several eclipse observing expeditions to South Africa, Libya, and China to test a new instrument for measuring the temperature and flow speed of the solar wind in the low corona .

Dr. Davila is also the initiator and chief organizer of the International Heliophysical Year program. Through this program a number of international distributed instrument observatories have been developed in coordination with international funding agencies, scientists, and the United Nations.

Coronal image taken by the NASA team during the Aug 2008 eclipse from an observing site in the Gobi desert.

Brian Dennis

Dr. Dennis is a solar astrophysicist, the RHESSI Mission Scientist, and either a Co-I or PI on three other projects:
- VxO for Heliophysics Data—Extending the Virtual Solar Observatory to Incorporate Data Analysis Capabilities
- Low-Cost Access to Space—Imaging X-ray Polarimeter for Solar Flares
- Facilitating the Joint Analysis of GLAST Solar Flare Observations

In addition, Dr. Dennis has been working on the analysis of RHESSI observations with the specific goals of optimizing the X-ray imaging and spectroscopy capabilities.

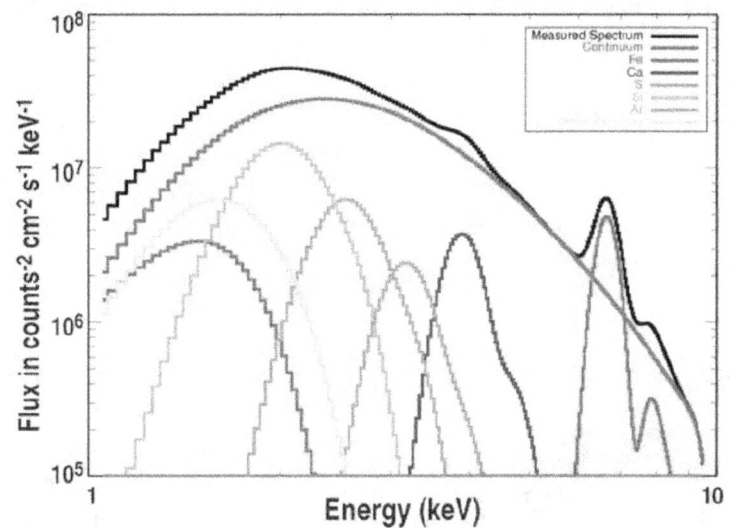

Contributions to the MESSENGER XRS count-rate spectrum (black histogram) of the thermal continuum emission (red) and the line fluxes from different elements (different colored curves). The photon spectra for these different components given by CHIANTI (v. 5.2) using the parameters that gave the best fit to the XRS spectrum were folded through the XRS response matrix to give the indicated colored curves.

Thomas L. Duvall, Jr.

Dr. Duvall, a solar physicist, has done research at GSFC for many years. His research is on understanding the solar interior using the techniques of helioseismology. During FY08, his research focused on near-surface phenomena, namely supergranulation, sunspots, and flows. All of this work uses the technique he pioneered—time-distance helioseismology—in which travel times for waves between surface locations are used to infer properties of the subsurface. New determinations of the lifetime (1.6 days) and size (27.1 Mm) of supergranules were made using a segmentation procedure on divergence maps derived from time-distance measurements. Preliminary work was done comparing magnetohydrodynamic (MHD) models of sunspots with observations of the travel times of f-mode waves. Reasonable agreement was obtained between a sunspot model with 3 kG magnetic field and f-mode travel times.

Timothy E. Eastman

Timothy Eastman, a space plasma physicist with Wyle Information Systems, LLC, joined the SPDF in 2002. Known for discoveries of the low-latitude boundary layer (LLBL) and gyro-phase-bunched plasmas, his research interests have included space plasmas, magnetospheric physics, plasma applications, data systems, and philosophy. He developed key foundations at NASA Headquarters and the National Science Foundation for major international and interagency projects including the International Solar Terrestrial Physics program, the interagency Space Weather Program, and the Basic Plasma Science and Engineering program. During 2008, his primary work was with the Science Proposal Support Office (SPSO, Code 605) to convene and implement 60 Red Team reviews for scientists in all Goddard Earth and space science areas, including space physics. These reviews have helped many proposers to increase their probability of success. In addition to managing the Space and Earth Science Data Analysis (SESDA)-II science support for SPDF, Dr. Eastman and his colleagues were successful this past year in obtaining grant support for International Sun–Earth Explorer (ISEE) Satellite Data Upgrades.

Justin Edmondson

Justin is a student from the University of Michigan doing doctoral work in theoretical solar astrophysics under Dr. Spiro K. Antiochos. His research includes theoretical work on the geometric and topological evolution of dynamic, three-dimensional structures in the global solar coronal magnetic field, the theory of three-dimensional magnetic reconnection, magnetized plasma instabilities, and the consequences thereof in the heliosphere.

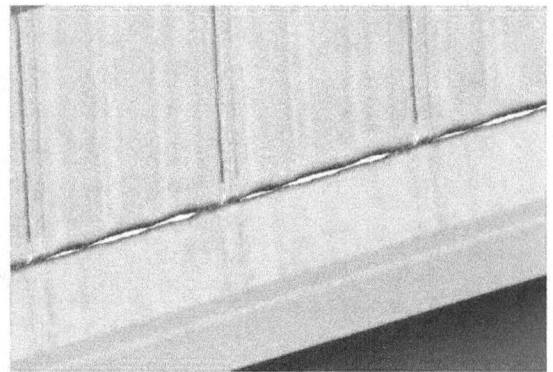

Several wavelengths of an interchange instability forming across a thin current sheet. The interchange mode occurs perpendicular to the plane containing the magnetic field lines. This instability offers a self-consistent mechanism that dynamically pinches anti-parallel field lines together across a thin current sheet, therefore initiating local regions of reconnection and driving up the reconnection rate of the system.

Don Fairfield

A magnetospheric physicist and Geotail Project Scientist, Dr. Fairfield retired and assumed emeritus status last June after 42 years at Goddard. He has been extending work done in 1985, which showed that nearly uniform precipitation of electrons over Earth's polar caps (polar rain) is caused by the direct entry of field-aligned solar wind "strahl" electrons along open magnetotail field lines. Recent work comparing interplanetary measurements with polar cap precipitation shows that small gradients in polar rain are due to field aligned potentials that are larger on field lines that connect to the interplanetary magnetic field further back in the tail. Dr. Fairfield helped man the Heliospheric Physics tent at the 2007 Folklife Festival on the DC Mall last July. This work was presented at the Fall 2007 AGU meeting and an extended version will be presented as an invited paper at the Fall 2008 meeting.

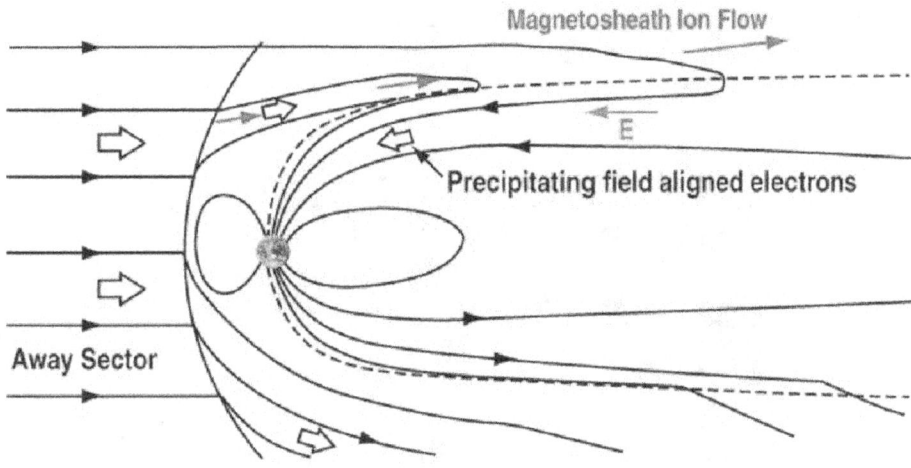

Solar wind electrons (open arrows) follow field lines down to the polar cap, but polar wind protons with greater tail-ward momentum (red arrows) cannot follow the electrons. To avoid a charge imbalance, a field-aligned electric field (blue arrow) is created that reduces the energy of incoming electrons, especially on field lines that go farther down the tail.

Artem Feofilov

Dr. Feofilov is an atmospheric physicist working at The Catholic University of America (CUA) at GSFC, Code 674. He has been doing research aimed at better understanding the fundamental processes governing the energetics, chemistry, dynamics, and transport of the mesosphere/lower thermosphere. His main efforts have been focused on improving water vapor retrieval from SABER[*]/TIMED measurements. Working with Drs. Kutepov, Goldberg, and Pesnell at GSFC he coupled a non-Low Thermodynamic Equilibrium (LTE) H_2O model with an extended model of O_2, O_3 photolysis products and used it as a reference for the SABER H_2O operational retrieval. The model was validated using the comparisons with ACE-Fourier Transform Spectrometer (FTS) occultation measurements. The first H_2O retrievals from SABER—6.3 μm radiance in the mesosphere demonstrate both qualitative and quantitative agreement with other experiments and simulations. The approach suggested by Dr. Feofilov has been accepted for the operational H_2O retrieval in the next release of SABER data.

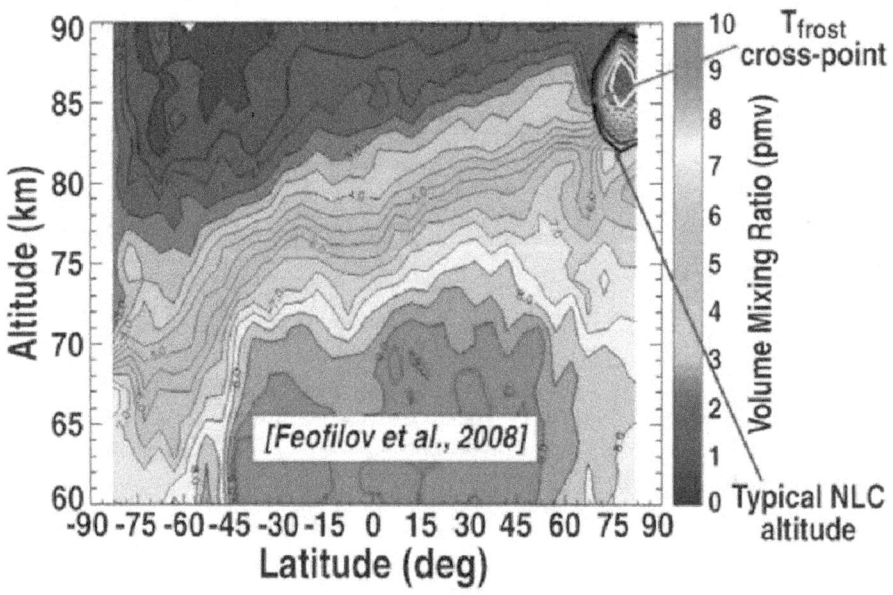

H_2O meridional distribution for 2004, day 197, retrieved from SABER/TIMED measurements

---

[*] SABER: Sounding of the Atmosphere using Broad band Emission Radiometry

Mei-Ching Fok

An astrophysicist in the Geospace Physics Laboratory. Dr. Fok has been working on modeling the inner magnetosphere, ionosphere and coupling between plasma populations. She has developed two kinetic models: the Radiation Belt Environment (RBE) model and the Comprehensive Ring Current Model (CRCM). In the past year, she have included wave-particle interactions in the RBE model. She improved the stability and reliability of the plasmasphere model, which is embedded in both the RBE and CRCM models. Efforts have been devoted to couple Dr. Fok's inner magnetosphere models with global magnetospheric models. Dr. Fok and researchers at the University of Michigan integrated the RBE model into Michigan's Space Weather Modeling Framework. Furthermore, through collaboration with scientists at the University of New Hampshire, the CRCM has been coupled to the OpenGGCM* MHD model. All these activities have yielded fruitful results and publications in journals.

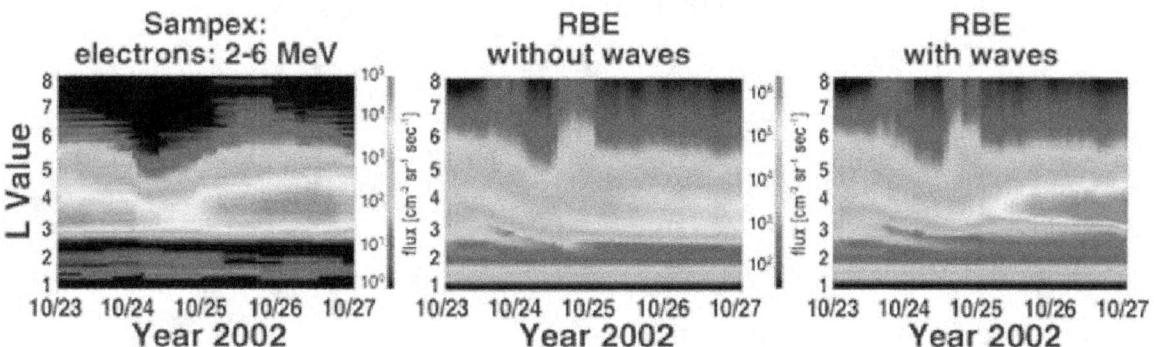

The L-time diagram of SAMPEX 2–6 MeV electron flux (left) during the magnetic storm on 2002 October 23–27 and RBE simulation without (middle) and with (right) inclusion of wave-particle interactions. It is obvious in this case that wave-particle interactions are necessary to produce flux enhancement during this storm.

---

* OpenGGCM: Open Geospace General Circulation Model

Shing F. Fung

Dr. Fung, a space scientist in Code 673, continued his research relating to global magnetospheric configurations. In a paper titled "Specification of multiple geomagnetic responses to variable solar wind and IMF input," published in March 2008, Fung and Shao (of the University of Maryland, College Park) demonstrated that magnetospheric states can be prescribed by corresponding *time-shifted* solar wind, IMF, and the multi-scale geomagnetic (Kp, Dst, and AE) parameters as shown by the blue lines in the accompanying figure. The solid and dashed (blues) lines mark the time progression or evolution of magnetospheric states. Using magnetospheric state prescriptions, the set of magnetospheric input parameters (top four panels) can be used to specify or "predict" the corresponding multiple geomagnetic responses (red curves) in the bottom three panels *simultaneously*. It is interesting to note that the magnetospheric state technique can correctly predict different observed Dst and AE conditions (black curves) even though the Kp conditions are quite similar (see the different sub-intervals enclosed by the solid and dashed blue lines). In addition to the 2-day interplanetary shock interval as shown, the same magnetospheric state technique was also used successfully to obtain out-of-sample predictions of geomagnetic responses in a 10-day multiple-storm interval and in the entire year of 2002 with similar predictive efficiency.

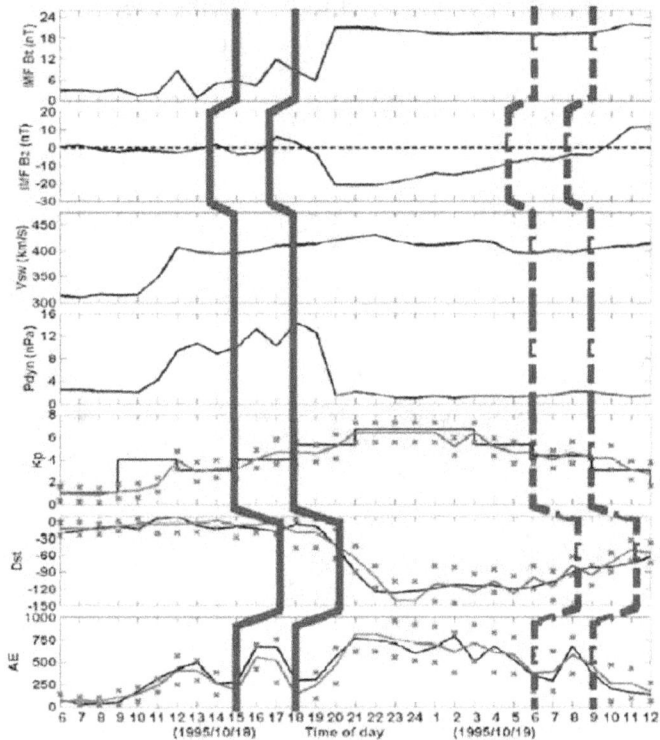

Magnetospheric states are prescribed by time-shifted solar wind, IMF, and the multiscale geomagnetic parameters as indicated by the blue lines. Successive blue (solid and dashed) lines represent the evolution of magnetospheric states. The bottom three panels show good agreements between the observed (black) and prescribed (red) geomagnetic responses. Predicted by the magnetospheric state technique. (Fung and Shao, 2008).

Dr. Fung is the recipient of a number of awards including the NASA Group Achievement Award and a Special Act Award.

Leonard Garcia

Mr. Garcia is a data acquisition scientist with Wyle Information Systems, LLC, who supports Code 670 through the SPDF. He supports the Satellite Situation Center (SSCWeb) through the maintenance of orbital data for more than a dozen current and past heliophysics missions, and is the primary point of contact for IMAGE satellite data held by SPDF. He is also a Co-I on the Virtual Wave Observatory (VWO, Shing Fung, PI). He has analyzed IMAGE Radio Plasma Imager data to study the evolution and spatial extent of plasmaspheric plumes. He also has participated in several Education and Outreach projects including the Solar System Radio Explorer Kiosk (SSREK) that he created, which teaches museum visitors about solar, terrestrial, and planetary radio emissions. He is also newsletter editor and archivist for the Radio Jove project, which teaches about planetary and solar radio emissions through the construction and operation of a simple radio telescope. In addition, he wrote an educational article on Space Weather for the Coalition for Plasma Science and participated in the Montgomery County Heritage Days 2008 at the site of the discovery of planetary radio emissions from Jupiter near Seneca, Maryland.

Left: Leonard Garcia at the Planetary Radio Emissions Discovery Site, Montgomery County Heritage Days 2008. Right: Mr. Garcia presenting SSREK at the Annapolis State House, 2008.

Holly Gilbert

HSD's new Associate Director for Science, Dr. Gilbert joined Code 670 in June 2008 where she has continued doing research on solar surface phenomena associated with Coronal Mass Ejections (CMEs). Specifically, she has recently published a paper investigating the interaction between global chromospheric waves and oscillating filaments and has just completed a study to determine the relative Hydrogen and Helium abundance in solar filaments. In addition to conducting research and leading the E/PO efforts of Code 670, Dr. Gilbert also served as a Working Group Leader at two meetings (SHINE* and SECCHI) and sits on the Solar Physics Division Committee and the Hale/Harvey Prize SPD Committee. She continues collaborations with Rice University (her previous institution prior to coming to GSFC) and supervises one Ph.D. student there.

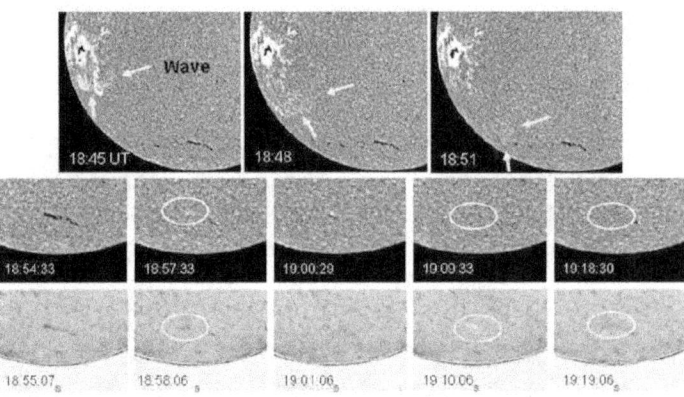

Wave observed in Hα data from the Mauna Loa Solar Observatory (MLSO) (top panel), and the initial stages of filament response to the passing wave in Hα (middle panel) and He I (λ10830) intensity (bottom panel) data from MLSO. White circles show the largest differences in appearance in the filament in the two lines.

Alex Glocer

Dr. Glocer joined Code 670 as an NPP in September 2008 after completing his Ph.D. in space and planetary physics at the University of Michigan. He has been doing research at GSFC with Dr. Mei-Ching Fok on modeling radiation belt electrons. Working with Dr. Fok, he has coupled the Radiation Belt Environment (RBE) model into the Space Weather Modeling Framework (SWMF) allowing for the ability to study the radiation belt population as a part of the space environment system; the initial results of this work have been submitted for publication. He also presented modeling results regarding the effect of ion outflow on magnetospheric composition and dynamics at the AGU spring meeting in Fort Lauderdale, Florida.

---

* SHINE: Solar, Heliospheric, and Interplanetary Environment

Melvyn Goldstein

Dr. Goldstein is a space plasma physicist who has been at Goddard since 1972, first as a National Research Council Postdoctoral Associate and, since 1974, as a member of what is now the Geospace Science Laboratory. His research focuses on a variety of nonlinear plasma processes that can be elucidated using data from the four Cluster spacecraft. In addition, Dr. Goldstein has participated in large and complex simulations of the origin of magnetohydrodynamic turbulence in the solar wind. He also serves as the Project Scientist for the Magnetospheric Multiscale missions and as NASA's Project Scientist for the ESA/NASA Cluster mission.

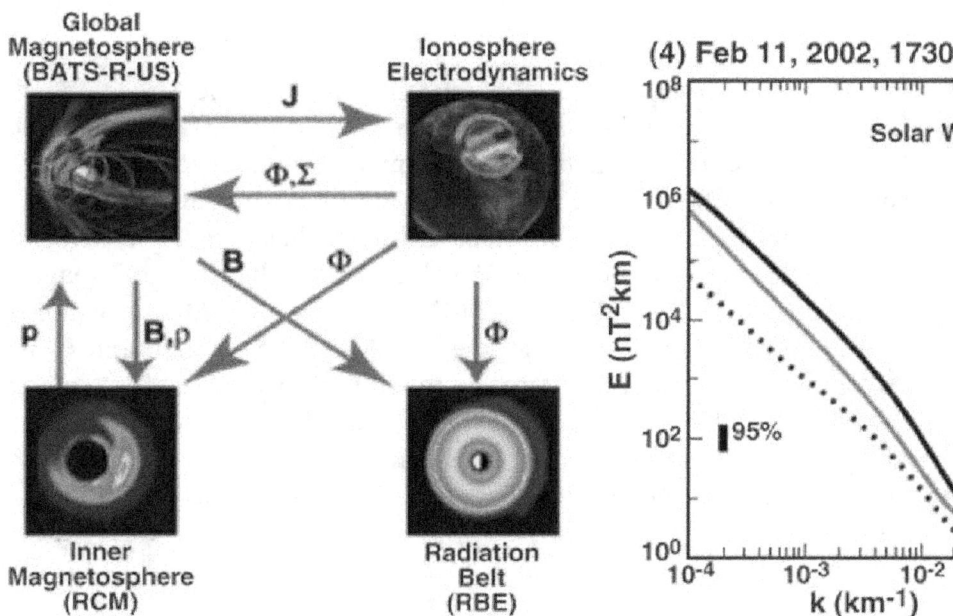

A schematic showing how the RBE model is coupled to other codes in the SWMF

A cut of the wave number spectra for two-dimensional (dark black), Alfvénic, (grey), and compressible (dotted) magnetic fluctuations in the solar wind as measured by the four Cluster spacecraft. The construction uses the "wave telescope" technique to determine the wave numbers directly.

Joe Gurman

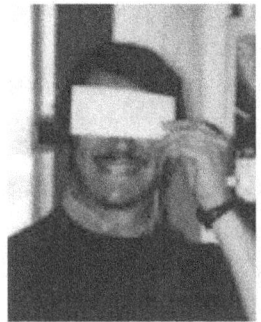

Dr. Gurman is the project scientist for both the SOHO and TRACE missions. He made presentations to the Senior Review of Heliophysics Operating Missions, the Heliophysics Data and Computing Working Group, and the symposium held to celebrate the tenth anniversary of the SOHO recovery. He also gave presentations at both the fall and spring AGU meetings.

Michael Hesse

Dr. Hesse, a space plasma physicist, is Chief of the Space Weather Laboratory (Code 674), and Director of the CCMC. His research focuses on the development and assessment of space weather modeling capabilities, and on basic research of the properties and dynamics of space plasmas. In his role as Lead Co-I for Theory and Modeling for NASA's MMS mission, he develops new theories of magnetic reconnection, and he advises the MMS project on MMS measurement priorities. As Director of the CCMC, he collaborates with governmental, academic, and commercial space weather interests across the globe. During 2008, Dr. Hesse gave nine invited talks on science and space weather topics. During the same period, he published nine papers in refereed journals, two of which as first author, and five as second author. In addition, he gave two short courses to students at the University of Texas.

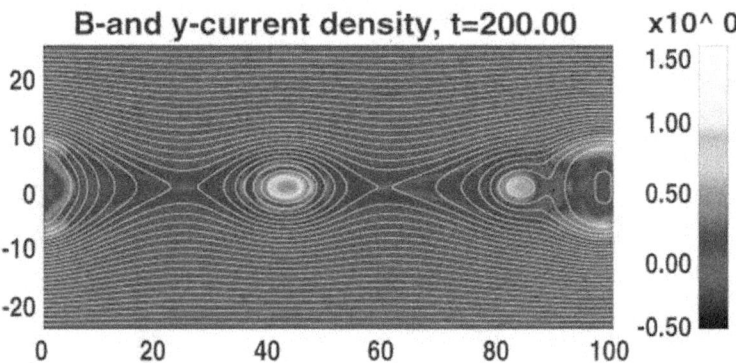

Magnetic field and current density during reconnection dynamics in a relativistic plasma. Modern research shows that nonrelativistic reconnection mechanisms carry over to relativistic plasmas.

Gordon Holman

A solar physicist, Dr. Holman primarily works on the analysis and interpretation of data from RHESSI. His scientific work has largely focused on obtaining a better understanding of energy release and particle acceleration in solar flares. He is the lead author on a review of the implications of X-ray observations for electron acceleration and propagation in solar flares, and coauthor of another related review paper, both submitted to *Space Science Reviews*. He was the Main Scientific Organizer for a three-day event at the 37th COSPAR Scientific Assembly (July 2008) in Montreal, Canada, titled "Magnetic Reconnection and Particle Acceleration in Solar Eruptions." In 2007, he received a NASA/Goddard Special Achievement (Peer) Award for his April 2006, *Scientific American* article "The Mysterious Origins of Solar Flares," and the 2007 Popular Writing Award for this article from the Solar Physics Division (SPD) of the American Astronomical Society (AAS). He served on the AAS/SPD Popular Writing Award Committee in 2008.

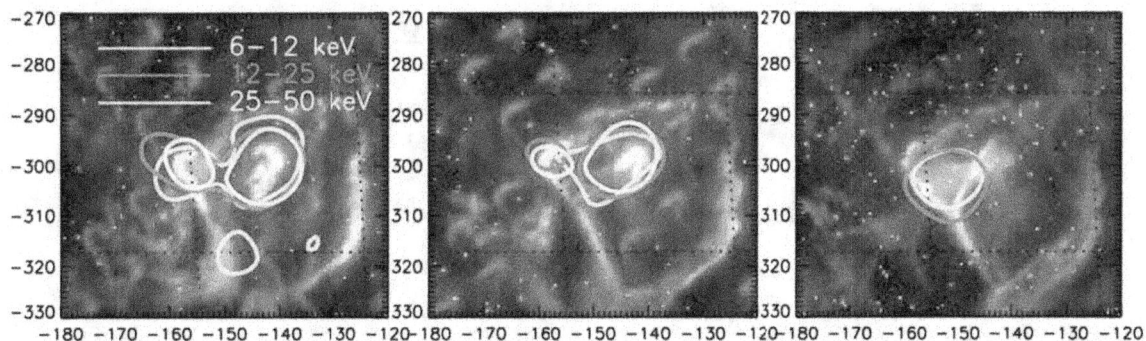

Three consecutive TRACE images with RHESSI contours of a flare on 2002 June 2, indicating primary electron acceleration in the solar corona before a wishbone structure is visible in the final TRACE image (Figure 3 of Sui, Holman, and Dennis 2008). The cusp of the wishbone (pointing downward in the figure, outward on the Sun) is thought to be associated with a coronal current sheet where magnetic reconnection occurs.

Joseph Hourclé

As the Principal Software Engineer with Wyle IS, Mr. Hourclé has been working on improvements to, and maintenance of, the Virtual Solar Observatory (VSO), as well as programming tasks for STEREO/COR1 and testing for support of SDO.

He gave an invited talk at the National Solar Observatory on design considerations for data, event, and feature catalogs, as well as other talks about the work being done at the VSO to the AAS SPD/AGU Joint Assembly pertaining to the work with catalogs, and a more general state of the VSO report at the Heliophysics Data and Modeling Consortium meeting. He had a paper accepted for publication and gave multiple presentations on applying Functional Requirements for Bibliographic Data

(FRBR), a reference model for library cataloging to scientific data, to disambiguate between the similar holdings held by an archive or across multiple data archives.

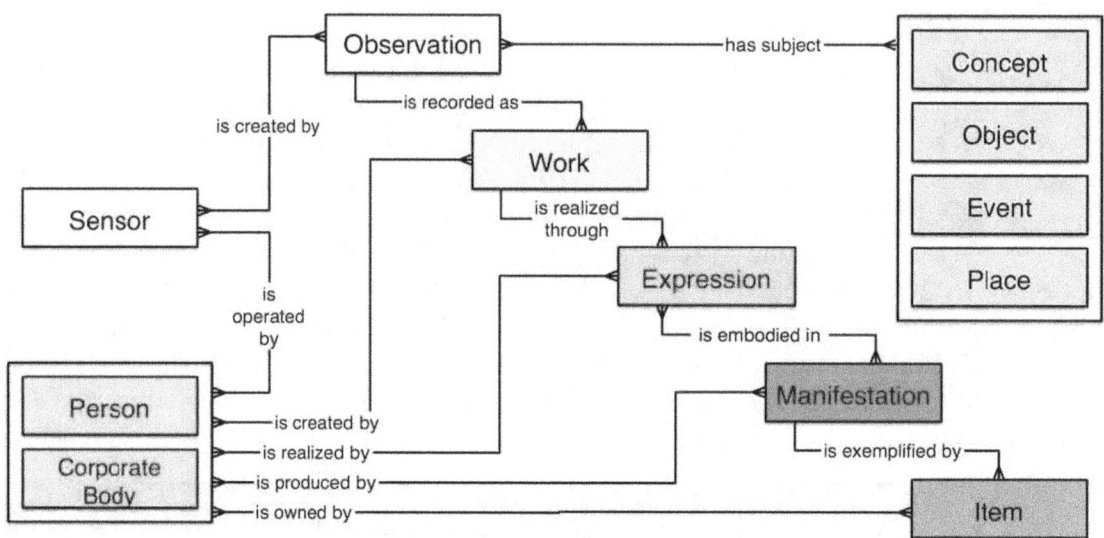

FRBR applied to scientific data

Kyoung-Joo Hwang

A Research Associate, Kyoung-Joo Hwang (UMBC), joined Code 673 in October 2008 has been doing research at GSFC with Dr. Goldstein on MMS pre-launch studies. Her recent computational study, continued her previous work at LASP, Univ. of Colorado, investigated the development and evolution of the parallel electric field (often structured as a localized double layer (DL)), electron phase-space holes, and particle heating, which are prevalent phenomena, and of long-standing interest in magnetospheric plasma, including the reconnection region. A simulation also focused on the feedback between a moving DL and particles, suggesting a new self-consistent model. This work is to be submitted to a science journal.

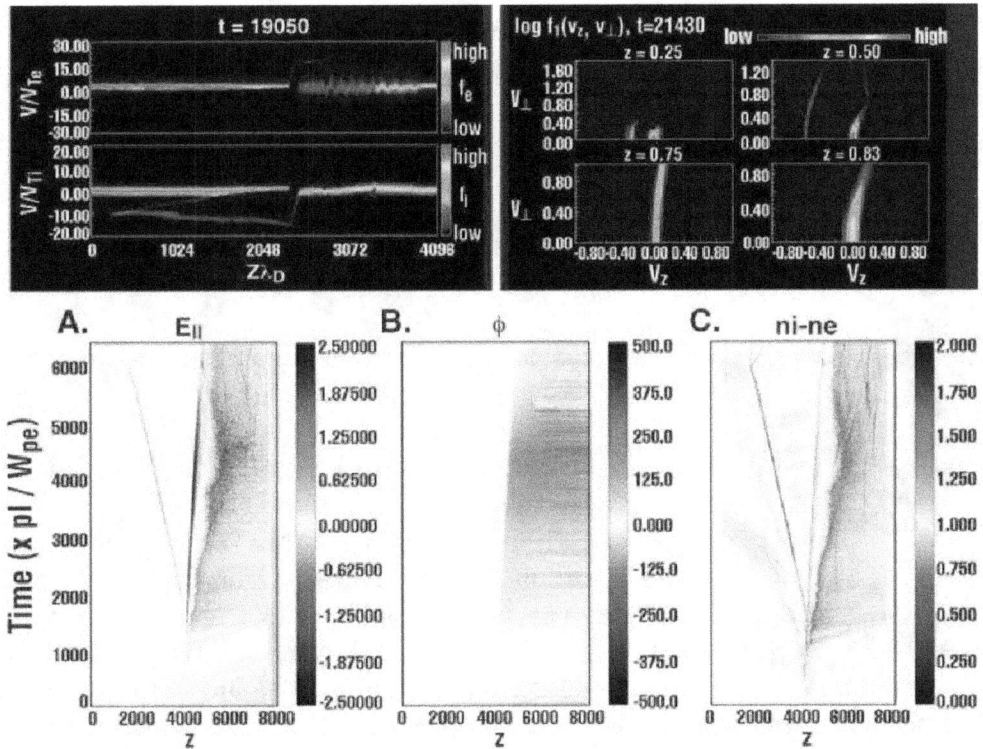

a) Electron (upper) and ion (lower) phase-space distributions along a simulation length, z (DL is located at $3/5 \times L_{sim}$, ripples in $f_e(z, v_z)$ indicate electron phase-space holes); b) 2-D velocity distribution function of ions, $f_i(v_z, v_y)$ at four specific locations ($1/4 \times L_{sim}$, $1/2 \times L_{sim}$, $3/4 \times L_{sim}$, and $5/6 \times L_{sim}$, note the formation of ion conics); and c) time history plots of $E_{||}$, potential, and current along the simulation domain, z (x-axis) and time (y-axis).

Jack Ireland

Dr. Ireland works for ADNET Systems, Inc., and in 2007–2008, his work lay in three primary areas. First, he worked in characterizing and understanding the magnetic complexity of active regions and its relation to solar activity. The second main focus of his research work in this period was the understanding of wave propagation in the solar atmosphere. The final research area was the development of user-friendly Web interfaces for the exploration of solar data sets, catalogs, and science. He expects to continue working in these research areas (and add a couple more) in the coming year. Dr. Ireland also supervised a summer student this year and undertook SOHO-EIT and Hinode Solar Optical Telescope (SOT) planning duties.

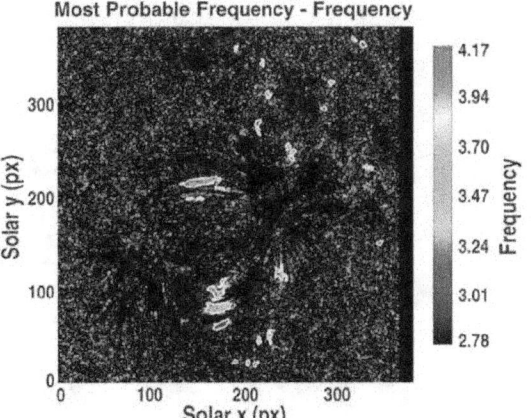

A map of oscillating material in TRACE 171 angstrom data (1998 July 14) found by an automatic detection algorithm recently developed by Dr. Ireland and co-workers.

Shaela Jones

Shaela Jones is a graduate student in physics at the University of Maryland. She is conducting thesis research at GSFC concerning the solar wind in the inner corona. Recently, she has given talks and posters at several conferences and submitted an article on the detection of so-called plasma "blobs" in the COR1 coronagraph aboard STEREO. Her other research interests include coronal tomography and CME initiation.

Mike Kaiser

Dr. Kaiser is a low-frequency radio astronomer in the Space Weather Laboratory. He serves as the Project Scientist for the STEREO Mission and is also the Deputy PI of the STEREO/WAVES investigation, PI of the Wind/WAVES investigations, and a Co-I of the Cassini/RPWS[*] investigation. His STEREO Project Scientist duties occupied nearly half of his time this year, including managing the hardbound edition of the STEREO instrument and mission description papers (21 papers) published in *Space Science Reviews*, and as Guest Editor of the Solar Physics Topical Issue concerning STEREO observations and analysis at solar minimum (70 manuscripts). His research during the past year has concentrated on the surprising detection by STEREO/WAVES of intermittent, and sometimes large quantities of, interplanetary dust of nanometer size impacting against the spacecraft. This work has also led to a reanalysis of the Cassini Jupiter flyby in late 2000 where similar particles of Jovian origin impacted the spacecraft and were detected by the RPWS instrument. He has also been involved analyzing Saturnian lightning discharges detected by the Cassini/RPWS instrument, a field of study he and his colleagues pioneered during the Voyager Saturn flybys nearly three decades ago.

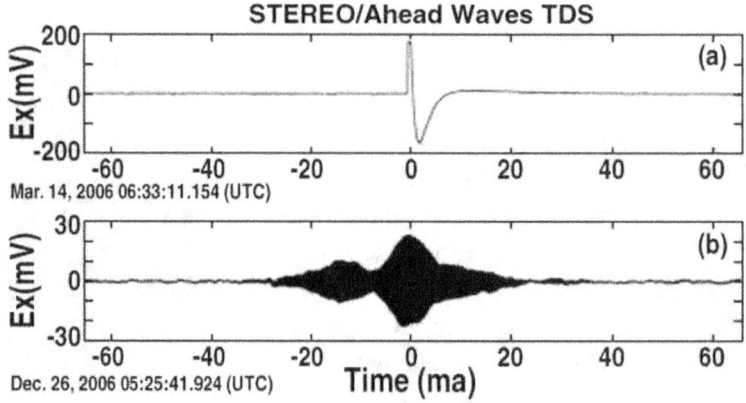

(a) Dust impact signature (b) compared with an in situ *Langmuir wave*

---

[*] Radio and Plasma Wave Science

George Khazanov

Prior to joining NASA in 2001, Dr. Khazanov was a fully-tenured Professor of Physics at the University of Alaska Fairbanks. Dr. Khazanov has extensive experience in space plasma physics and simulation of geophysical plasmas. His specific research areas include: analysis of hot plasma interactions with the thermal space plasma with special emphasis on hot plasma instabilities, investigation of current-produced magnetic field effects on current collection by a tether system, space plasma energization and transport, kinetic theory of superthermal electrons in the ionosphere and plasmasphere, hydrodynamic and kinetic theory of space plasma in the presence of wave activity, theoretical investigation and numerical modeling of ionosphere-plasmasphere interactions, theoretical study of artificial injection of charged and neutral particles into the ionosphere, and waves and beam-induced plasma instabilities in the ionosphere. Dr. Khazanov supervised and directed more than 30 M.S. and 15 Ph.D. graduates. He is the author or coauthor of 5 books and approximately 250 peer reviewed publications.

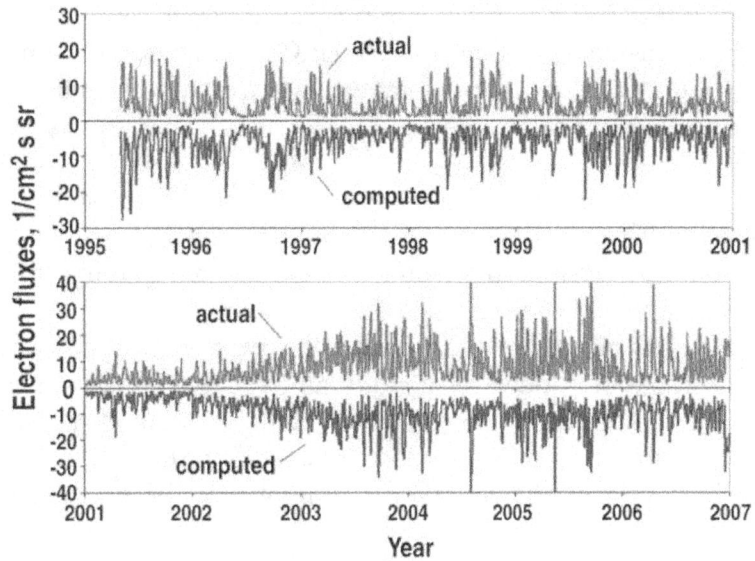

The daily averages of the cube root of actual and computed relativistic (>2 MeV) electron fluxes from 1995 through 2006.

Joe King

Dr. King is interested in long term solar wind variations and in the best approaches to shifting data from upstream of Earth's bow shock to Earth's more immediate vicinity for solar wind-magnetosphere coupling studies. He is active in creating value-added multi-source data sets, most notably the 1963–2008 solar wind data set called OMNI, and is currently creating, with N. Papitashvili, an integrated interface called HelioOMNIWeb to all GSFC/SPDF heliospheric data. In addition, Dr. King is active in the emerging Heliophysics data environment characterized by distributed data, VxOs, and metadata standards (SPASE). Formerly, he was the Head of the National Space Science Data Center (NSSDC) and was the Interplanetary Monitoring Platform (IMP) 8 Project Scientist. He is a past recipient of the AGU Edward Flynn Award.

Alex Klimas

Collisionless magnetic reconnection is an ubiquitous process that plays an important role in the dynamics of many space, laboratory, and astrophysical plasma systems. Because it is a complicated nonlinear process, much of what has been learned about collisionless reconnection has come from simulation studies. From these studies, a "Hall reconnection model" has emerged as the generally accepted paradigm for this process. Recently, however, simulations done with open, rather than the traditional periodic boundary conditions have challenged the Hall reconnection model. The purpose of the current, ongoing research is to modify an available electromagnetic 2 ½-dimensional particle-in-cell (PIC) reconnection simulation code to incorporate open boundaries and then to investigate this recent challenge to the Hall reconnection model using the modified code.

We have developed a new, and far more general, algorithm for open boundaries in PIC simulation codes. A simple method that allows passage of particles through an open boundary while maintaining a zero normal gradient for the particle distributions has been developed. The biggest difficulty with this construct is keeping track of particles lost and gained but this can be done with very little additional computational load. In addition, compatible boundary conditions have been imposed on the electromagnetic field at the open boundaries. We have achieved smooth passage of electromagnetic structures (magnetic islands, outflow jets, etc.) through the boundaries with no trace of charge buildup near the boundaries or anywhere else on the computational grid.

The newly available collisionless reconnection simulation code with open boundaries is performing well and is being used at present to investigate the recent challenge to the Hall reconnection paradigm.

Jim Klimchuk

Dr. Klimchuk joined the Solar Physics Lab in early January of 2008. He devoted most of his time to studying the heating of the solar corona, using a combination of theoretical models and observations from the TRACE and Hinode missions. He developed a highly efficient numerical simulation code (EBTEL) for investigating the hydro-dynamic response coronal plasmas to various types of heating. By comparing predicted and observed radiation

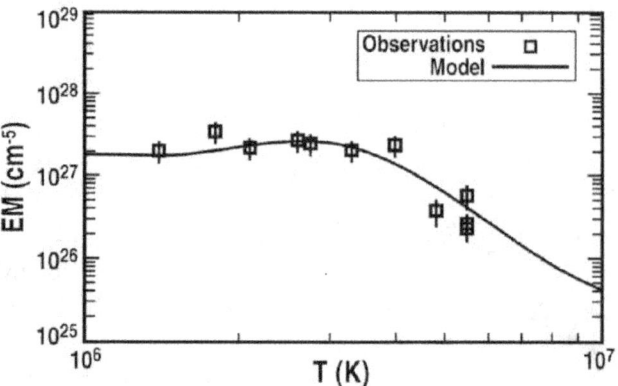

Emission Measure distribution (amount of plasma as a function of temperature) observed by the EIS spectrometer on Hinode (squares) and predicted by a model active region heated by nanoflares (curve).

signatures (EUV emission line intensities and profile shapes, as well as broad-band intensities in the EUV and soft X-ray), he showed that much of the corona is heated impulsively by nanoflares. He also used MHD simulations to demonstrate that the secondary instability of electric current sheets is the likely source of the nanoflares. Klimchuk remained active in community service as Vice-Chair of the AAS Solar Physics Division, President of IAU Commission 10, and member of the AAS Committee on Astronomy and Public Policy, Solar Physics Editorial Board, and GSFC Deputy Director's Council on Science. He also led the NASA LWS Focus Team on Solar Origins of Irradiance Variation.

Yoji Kondo

Dr. Kondo's research centers around the physical properties and evolutionary status of close binary stars. He is a Co-I on Kepler, an observatory to be launched in 2009 that will be the first mission capable of finding Earth-size planets around other stars. Earlier, he was the Project Scientist for the Internal Ultraviolet Explorer (IUE) and the Extreme Ultraviolet Explorer (EUVE). Dr. Kondo is also active in public outreach, writing, and speaking on various aspects of NASA's space science program, including the complementary roles of human and robotic exploration. He has edited 13 professional books and is the recipient of the NASA Medal for Exceptional Scientific Achievement, the National Space Club Science Award, and the Isaac Asimov Memorial Award.

Andriy Koval

Dr. Koval is a NASA Postdoctoral Program Fellow in Code 672. He has been doing research of interplanetary shock properties by improving the technique for determining shock local parameters from the Rankine-Hugoniot conservation equations. The modified technique simultaneously determines the shock normal direction and propagation speed leading to a more accurate solution. The results were presented at the 2008 AGU Joint Assembly. He has been also working on the improvement of the Wind Magnetic Fields Instrument (MFI) instrument calibration to eliminate the spin noise in the high resolution (0.1s) data.

Maxim Kramar

Dr. Kramar, of CUA, has been working on 3-D reconstruction of the electron density in the solar corona by the regularized tomography method, and based on data from STEREO spacecrafts. Because the solar corona is optically thin, coronal observations are essentially integrated over the line-of-sight (LOS). It is therefore impossible to resolve the structure of the corona along the LOS if observations are provided from a single view direction. Observations from different view positions are necessary to reconstruct the 3-D coronal structure and is the essence of the tomography inversion method.

When observations are only from a single view direction, a rigid rotation of the coronal density structures with the Sun about the ecliptic must be assumed in order to apply the tomography technique. As a consequence, only structures that are stationary over half a solar rotation can be reconstructed. The obtained electron density structure could be used for testing coronal models as the reconstruction reflects nearly real coronal density structure within assumptions described above. Results of this work have been described in a paper that was submitted for publication in *Solar Physics*.

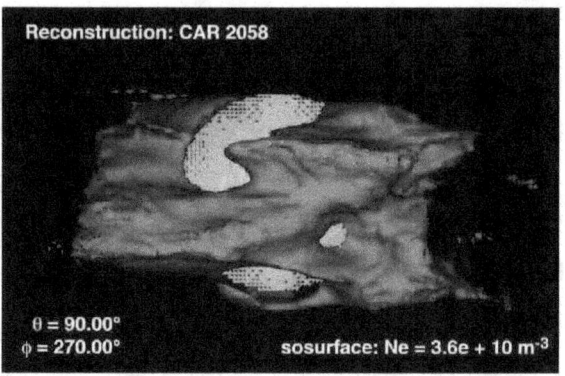

Isosurface of the electron density in the solar corona at value 3.6×10^{10} m^{-3} for the period of Carrington Rotation 2058. The orange inner sphere corresponds to distance of 1.5 solar radii.

Therese A. Kucera

Dr. Kucera is a solar physicist who joined Code 671 as an NRC postdoctoral fellow in 1993–1995, and then worked in the branch as a contractor until 2001 when she became a civil servant. She has been doing research into the solar atmosphere with special emphasis on solar prominences and ultraviolet spectroscopy. Her other interests include active regions and coronal cavities. She has served as the Deputy Project Scientist for STEREO and SOHO, and as the STEREO E/PO lead. She is currently on detail to NASA Headquarters as the Heliophysics Solar Discipline Scientist. She received a NASA Group Achievement award for the STEREO 3-D event.

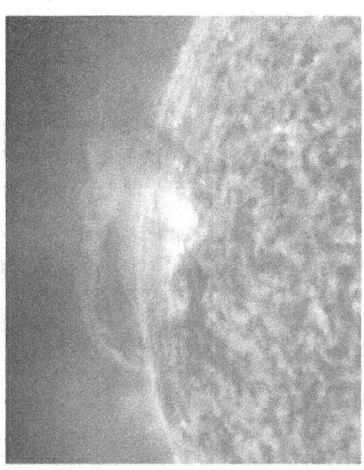

Solar prominence observed by SOHO/EIT at 304 Å.

Alexander Kutepov

Dr. Kutepov is an atmospheric physicist with the Department of Physics at the Catholic University of America (CUA). He performed (together with Dr. Feofilov at CUA and GSFC) the non-LTE analysis of broadband infrared (IR) (5.5–100 µm) limb observations of the Martian atmosphere by the Mars Global Surveyor (MGS) Thermal Emission Spectrometer (TES) bolometer. For this study, he employed the Accelerated Lambda Iterations for Atmospheric Radiation and Molecular Spectra (ALI-ARMS) non-LTE model and the forward fit algorithm developed and applied for retrieving mesospheric and lower thermospheric temperatures and trace gas densities from Earth's limb radiances. It led to the first global retrievals of pressures and temperatures in the middle and upper Martian atmosphere (60–95 km) from the MGS/TES observations, which demonstrate new interesting features associated with various forms of the dynamic activity.

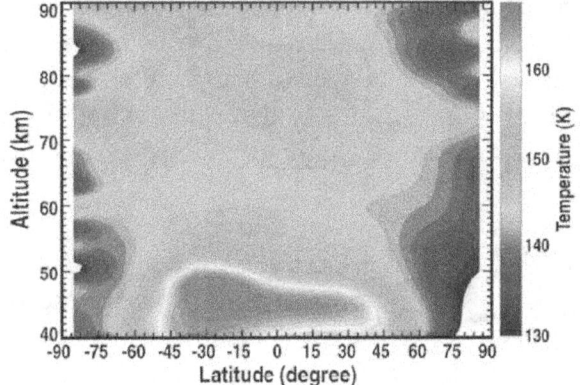

Nighttime Temperature Distribution in the Martian atmosphere for s=0, retrieved from the MGS/TES Bolometer data

Together with Drs. Feofilov, Goldberg, and Pesnell (all at GSFC). Dr. Kutepov also has been working on the water vapor density retrievals in Earth's mesosphere and lower thermosphere from the IR broadband SABER/TIMED observations of Earth's limb. The first H_2O retrievals from SABER 6.3 µm radiance in the mesosphere demonstrated both qualitative and quantitative agreement with other observations and model simulations. They are used for validating the SABER operational H_2O retrievals performed for the next release of the SABER data.

Nand Lal

Dr. Lal is a member of the Voyager Cosmic Ray Subsystem (CRS) team. Highlights of the CRS team's accomplishments in the past year include: contribution to the successful Senior Review proposal for the Voyager Interstellar Mission, and publication of observations from Voyager-2's crossing of the termination shock: "An asymmetric solar wind termination shock" (published in *Nature*). In the CRS team's continuing effort to facilitate broader use of CRS observations, Dr. Lal has developed tools that make it easier to use CRS measurements with other charged particle observations. As a member of the Virtual Energetic Particle Observatory (VEPO), he has been actively involved in efforts to improve usability of energetic particle observations by the broader science community. He developed the metadata descriptions required for access to CRS observations through VEPO and the Virtual Heliospheric Observatory.

Derek Lamb

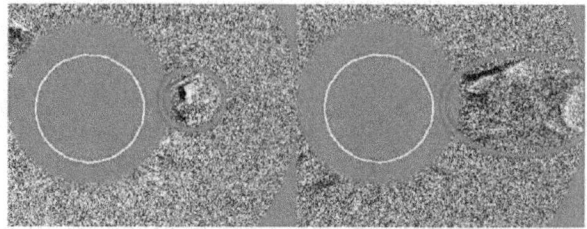

After defending his thesis at the University of Colorado, Dr. Lamb (Catholic University of America) started at Goddard in November. He is analyzing solar coronal mass ejections as observed by the STEREO-SECCHI/COR1 coronagraphs. Some of these ejections show a particularly slow liftoff, and he is analyzing these events with multiple instruments from multiple observing positions to determine whether or not these events are fundamentally different from their more common, much faster counterparts.

Running difference images of the corona as observed by STEREO/COR1. (left) Red circle highlights the initial appearance of a CME. (right) Red ellipse highlights significant structure still visible over 8 hours later.

Guan Le

Dr. Le is a magnetospheric physicist in the Space Weather Laboratory (Code 674). She is the science lead for the Space Technology 5 (ST5) mission and has been studying field-aligned currents (FACs) using multi-points magnetic field measurements from ST5. Her research focuses on the temporal variations of FACs in short time scales, as well as the ionospheric closure currents of FACs. She is a science Co-I of the Vector Electric Field Instrument on the C/NOFS mission. Her research focuses on the ionospheric currents using the C/NOFS magnetic field data. She is the Deputy Project Scientist for MMS and the Project Scientist for Geotail. She is a member of AGU's Space Physics and Aeronomy (SPA) Section executive committee, and serves as the editor of SPA Web site and co-editor of the SPA Newsletter.

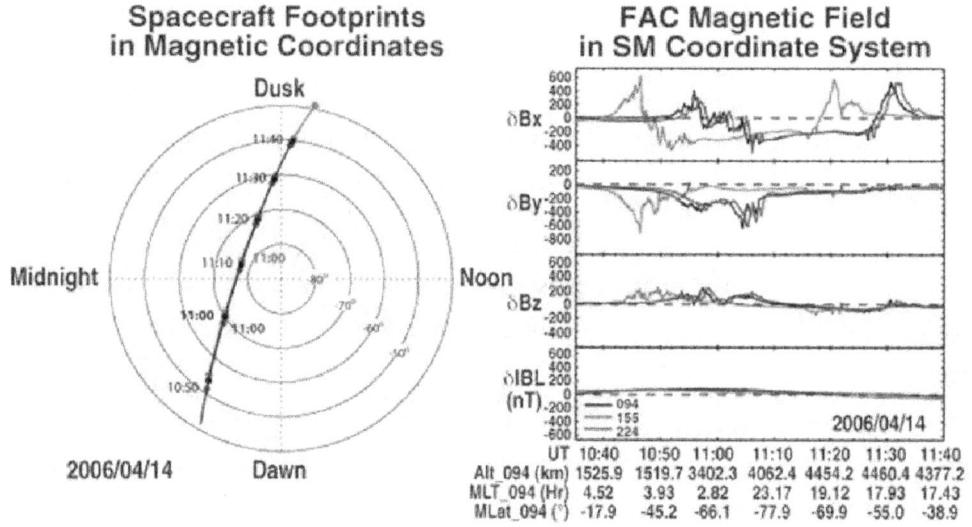

Ronald P. Lepping

Interplanetary magnetic clouds (MCs) are known to cause some of the most intense geomagnetic storms. The north-to-south type of magnetic cloud is expected to be the most prevalent for the next six or seven years. Ron, in collaboration with C.-C. Wu and T. Narock, used this fact to develop a practical scheme to predict the latter part of the MC from the earlier part, in order to help forecast the expected geomagnetic storm many hours in advance, with an extra hour lead-time because the spacecraft is expected to be at L1 (e.g., ACE). The "Comprehensive" MC fitting program differs from previous, and most other, MC fitting programs, in that it maintains the same estimate of the MC's axis from the older fit-program, but attempts to account for MC expansion and compression explicitly. The previous fit program is usually not accurate enough to estimate the intensity of the MC's field, but it is usually very accurate in obtaining the MC's axis and useful in helping to affirm if an MC is really being observed in the first place, early in the scheme. It is expected that this program will predict storms by five or six hours before actual occurrence time, for most MCs.

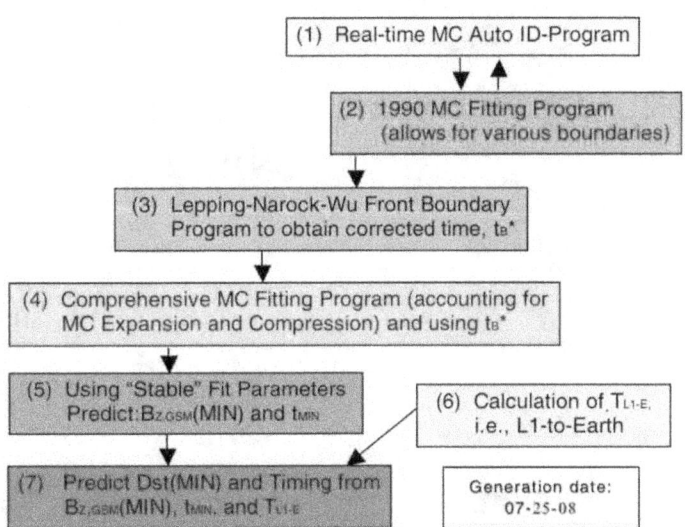

Real-Time Dst forecasting scheme based on MC observations and parameter modeling.

Alexander S. Lipatov

Dr. Lipatov, of UMBC, has a collaboration with Drs. E.C. Sittler, R.E. Hartle, and J.F. Cooper in 3-D hybrid-kinetic simulation of the plasma environment near Titan, Europa, and the Solar Probe. His basic research interests concern numerical simulation in astrophysical and laboratory plasmas:

- Development and application of the advanced multiscale simulation codes including the hybrid and the Complex Particle Kinetic concepts;
- Global multidimensional, multiscale, multifluid, and hybrid (fluid-kinetic) simulations of the interaction of the solar wind with the magnetospheres of Earth, Venus, Mars, comets, and moons;
- Hybrid/kinetic multiscale simulation of turbulent processes, particle heating and acceleration at the front of collisionless shocks, and magnetic field reconnection in the plasma systems with reversed magnetic field configuration, with application to solar flares, magnetosphere of the planets, bow shocks, interplanetary shocks, and the termination shock;

Hybrid simulation of the (plasma and dust) beam propagation in plasma;
Boltzmann (kinetic/fluid) global simulation of the interaction of atoms from the local interstellar medium with the heliosphere.

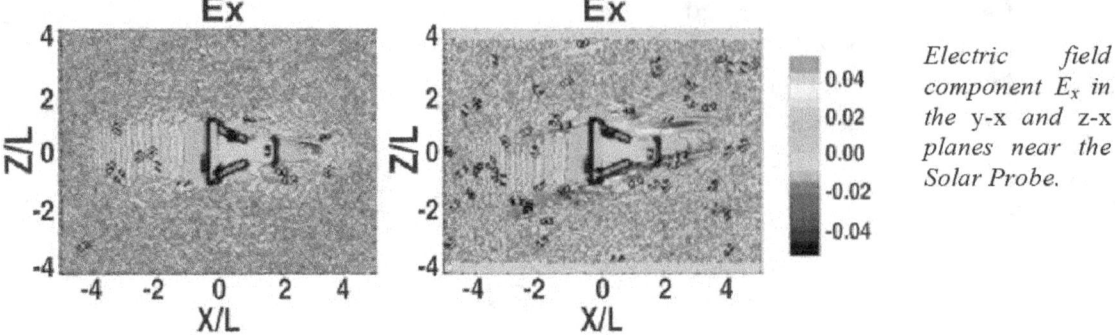

Electric field component E_x in the y-x and z-x planes near the Solar Probe.

Wei Liu

Dr. Liu is an NPP Fellow who joined Code 671 in January 2007. He has been working with Drs. Dennis and Holman in the GSFC RHESSI group. The objective of his research is to understand particle acceleration and energy release mechanisms in solar flares. His primary approach is analyzing soft and hard X-ray data obtained by RHESSI, together with complementary numerical modeling using a novel technique of combined particle and hydrodynamic simulation. His major findings in the past year include locating the magnetic reconnection site in an over-the-limb flare by analyzing RHESSI images and spectra of spatially resolved sources. He presented his research at several meetings including the 2007 AGU Fall Meeting and the 37th COSPAR Scientific Assembly. He has published papers in refereed journals and an extended version of his Ph.D. Thesis in a book.

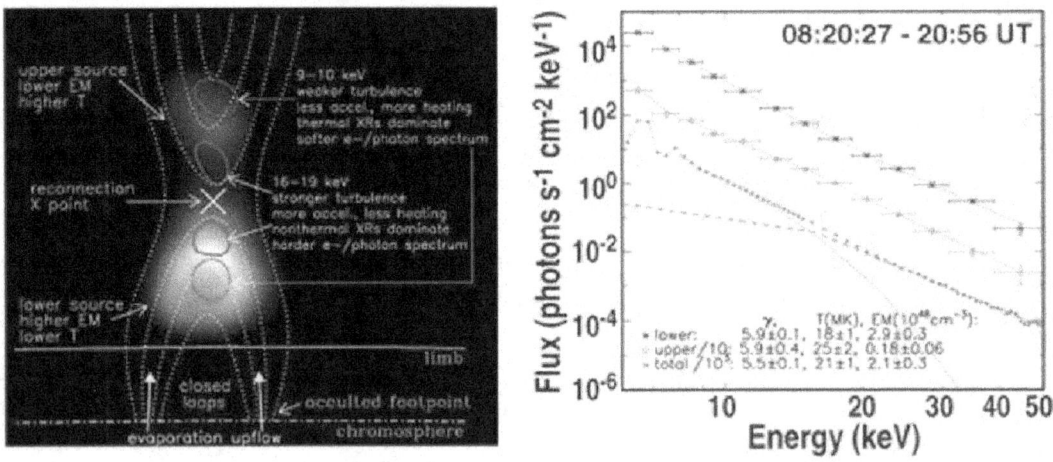

Left: *Sketch of the physical scenario for a solar flare superimposed on the RHESSI 14–16 keV image (green background), overlaid with the corresponding 9–10 (red) and 16–19 keV (blue) contours.* **Right:** *Spectra of the lower and upper coronal sources and the spatially integrated spectrum (labeled as "total").*

Robert McGuire

Bob McGuire is currently the Associate Director for Science Information Systems in Code 670. He also leads and directs the Space Physics Data Facility (SPDF) project as its Project Scientist. SPDF develops and operates multi-mission and active archive data and display services (e.g., CDAWeb), orbit planning and display services (e.g., SSCWeb), and supports the maintenance and use of the Common Data Format (CDF) heliophysics format standard, among other activities. He was the Lead Scientist for IMP-8, and the PI on the Goddard Medium Energy Experiment (GME) energetic particles experiment until the failure of IMP-8 in 2006. He also has science interests in solar and other heliospheric particle composition and acceleration, as well as data systems, and he is the Co-I on several current Virtual discipline Observatory (VxO) investigations.

Jan Merka

Dr. Merka is a space physicist with UMBC, and has been at GSFC since April 2001. He has been doing research in terrestrial bow shock properties and modeling. His 2008 activities were directed towards the development of the Virtual Magnetospheric Observatory, a portal for a unified search and retrieval of data for magnetospheric research (http://vmo.nasa.gov). Working with Dr. Szabo and T. Narock, he also works on the Virtual Heliospheric Observatory project dedicated to heliospheric data products. In 2008, he was either the prime author or co-author on 5 papers and 11 presentations at international meetings.

Ryan Milligan

Dr. Milligan is an NPP Fellow working with Dr. Brian Dennis as part of the RHESSI group. His research primarily focuses on solar flare observations by combining data from RHESSI and Hinode. His first publication from this work showed that microflares can achieve higher coronal temperatures if less energy is used in accelerating particles. This work was presented at the 37th COSPAR meeting in Montreal and the UK National Astronomy Meeting in Belfast, and was the subject of a NASA press release. He is continuing this work by investigating the process of chromospheric evaporation using the two spacecraft, and also the flare/CME relationship using RHESSI and STEREO. Dr. Milligan has hosted two students from Trinity College of Dublin this year who have been working on flare cooling using RHESSI and the SOHO Coronal Diagnostic Spectrometer (CDS) and statistical analysis of flare temperatures using 30 years of GOES data. Dr. Milligan is also one of the Max Millennium Chief Observers, has served on a NASA review panel, and helped out at the NASA exhibit at the 2008 Folklife Festival.

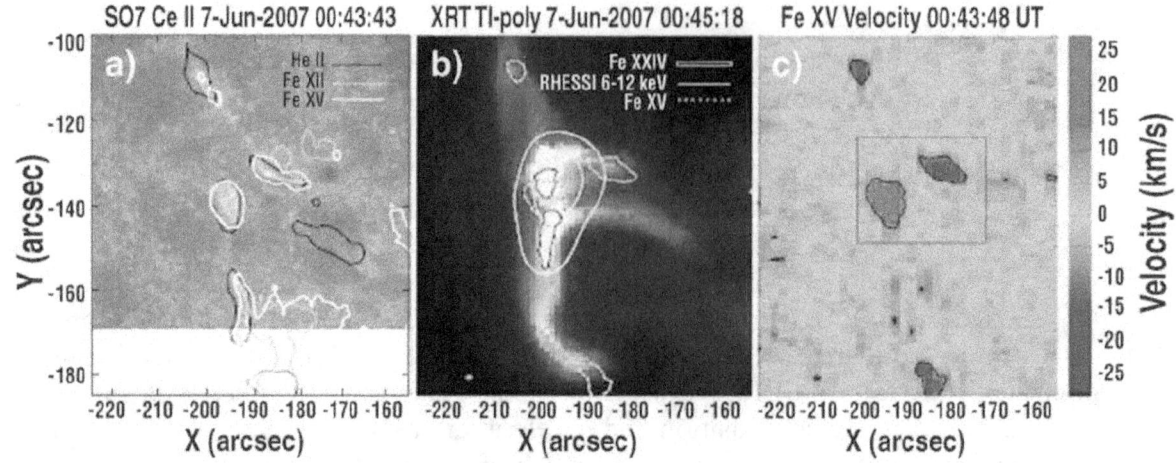

Images of a solar flare using each of the instruments on Hinode plus RHESSI. Left: Solar Optical Telescope (SOT) Ca II H image. Center: X-Ray Telescope (XRT) image with RHESSI contours overlaid. Right: an EUV Imaging Spectormeter (EIS) velocity map in Fe XV.

Tom E. Moore

Dr. Moore is HSD's Deputy Director. His principal research activities in FY08 were: Fast Plasma Investigation development; IBEX-Lo preparations for Mission Operations and Data Analysis (MO&DA); MO&DA activities for the Polar/TIDE[*] investigation; and an LWS Targeted Research and Technology (TR&T) task "Storm Time Plasma Redistribution: Processes and Consequences." The first two of these benefited from experience with the others, which came together in a study reporting the first magnetospheric simulation to include all relevant observed plasmas: solar wind, polar wind, auroral wind, and plasmaspheric wind. The newest element simulated was the plasmaspheric wind, created when reconnection with the solar wind magnetic field liberates plasmas usually trapped deep in the magnetosphere (the plasmasphere), circulating them to the magnetospheric boundary layers, where they are lost downstream, or re-circulated through the magnetotail to the inner magnetosphere with large increases in particle energy. The surprising result was that the plasmasphere makes a small contribution to the storm time plasma pressure inflating the inner magnetosphere, comparable to that of the polar wind. This work was the subject of two invited talks and one contributed talk, and was designated the most striking result to come out of the LWS TR&T-focused science team activity in 2008.

---

[*] TIDE: Thermal Ion Dynamics Experiment

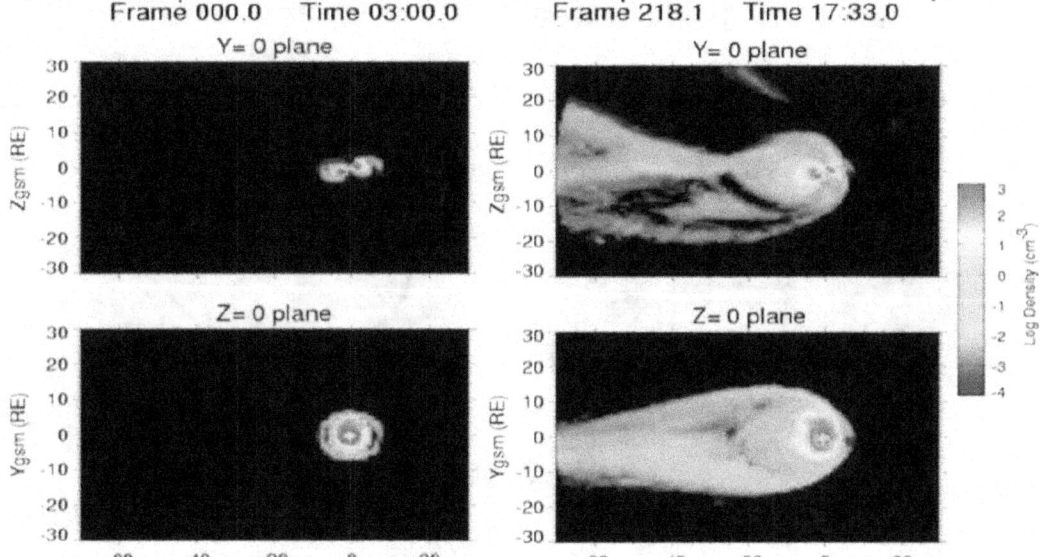

A comparison of the normal plasmasphere (left panels) with the circulating plasmasphere, which results from storm time circulation in the Earth's magnetosphere (right panels). On each side, the upper panel is a meridian plane and the lower panel is the equatorial plane. Solar wind enters the simulation from the right. Proton density is scaled according to the color bar.

Teresa Nieves-Chinchilla

Dr. Nieves-Chinchilla completed her Ph.D. at Alcala University of Madrid (Spain). Her dissertation work related to the study of the evolution of magnetic clouds in the interplanetary medium. She developed a non-force-free model (with expansion and deformation in the cloud cross-section) and its implementation in an algorithm to model Wind data.

She is currently an NPP fellow working with Dr. A.F.-Viñas. Some of her recent research has been related to the analysis and modeling of the kinetic thermodynamic properties of electrons inside magnetic clouds using Wind data from the Solar Wind Experiment (SWE) experiment. An example of the electron Velocity Distribution Function (VDF) modeling with kappa distribution is shown. Dr. Nieves-Chinchilla has developed an algorithm which has been implemented in an analysis and visualization tool (i.e., the Magnetic Cloud Analysis Tool, MCAT) for the systematic study of the magnetic field topology of magnetic clouds.

Currently, she has been working in the modeling of high time, energy, and angular resolution electron VDFs from the Cluster/PEACE[*] instrument (an example is shown in the figure following) to study the kinetic aspect of the electron distribution function. She

---

[*] PEACE: Plasma Electron and Current Experiment

is focusing on the strahl electron component to understand its role on processes that regulate plasma instabilities in the solar wind.

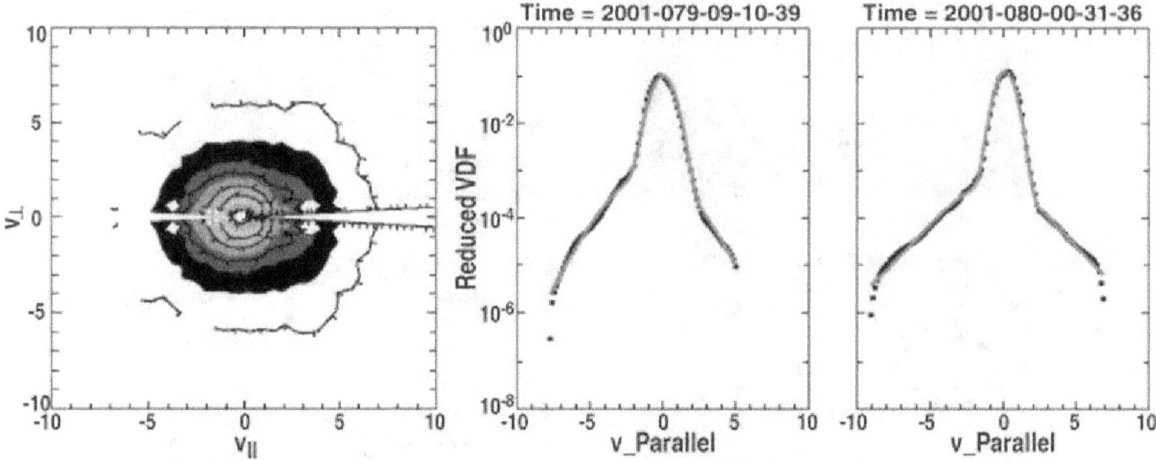

Sten Odenwald

Dr. Odenwald is an astronomer with ADNET/Catholic University. He has been involved with investigations of the cosmic infrared background since 1996, as is currently using data from the 2MASS survey with Dr. Alexander Kashlinsky. He is an active NASA educator, and is the Education Lead for both IMAGE and Hinode, as well as the Sun–Earth Connection Education Forum. This year he co-wrote, with Dr. Jim Green, an article for *Scientific American* on solar superstorms and a popular-level article on superstorms for *Astronomy* magazine. He will also be appearing on two *National Geographic* TV specials on space weather and solar storms. He gave the keynote talk at the Heliophysics Summer School in Boulder in July. His most recent research involves chronicling the historical, human impacts of space weather. In December he gave a popular-level talk at the Dudley Observatory on "21st Century Cosmology." He was recently awarded a 3-year NASA education grant to continue his "Space Math @ NASA" projec,t which develops K–12 math problems based on NASA mission data and science goals. Currently, there are over 1000 teacher participants in 50 states and over 30 countries, and 20 NASA missions and programs contributing math ideas across all NASA directorates.

Leon Ofman

Prof. Ofman, an Associate Research Professor at CUA, has been a solar physicist at GSFC since 1992. He is working on theoretical MHD models of waves in coronal loops and in active regions, in collaboration with other scientists at Goddard and around the world. Prof. Ofman's models are guided by SOHO, TRACE, and Hinode EIS observations. He was among the major contributors to the recently emergent field of coronal seismology—the use of wave observations in the solar corona for determining unknown coronal properties, such as magnetic field strength, density, and temperature. Prof. Ofman is working on kinetic models of solar wind plasma, using multifluid and hybrid codes to study heating and acceleration of the solar wind in the inner corona. Recently, Prof. Ofman served on NASA's Science and Technology Definition team for the Solar Probe+ mission, aimed at observing *in situ* the innermost part of the solar corona, where solar wind acceleration is taking place. Prof. Ofman is also involved in empirical modeling of solar wind and streamer structure, using observations to constrain the heating function, and the effective temperature of the solar wind.

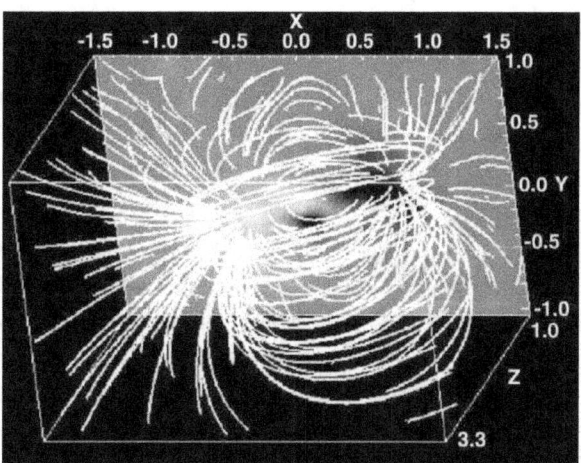

The 3D MHD model of an active region initialized with magnetic field extrapolated from the photosphere. A pulse simulating the effect of a flare introduced waves into the active region in the model. The study of the propagation of the waves enables determining the magnetic and plasma properties of the active region.

Keith Ogilvie

Dr. Ogilvie, as the PI of the PlasMag instrument on the Deep Space Climate ObserVatoRy (DSCOVR, formerly Triana) has been assisting in a study for NOAA on the cost and issues involved in launching the spacecraft, previously in storage here at GSFC, as an upstream, real-time monitor for space weather. In addition, Dr. Ogilvie has assisted with Wind Project Science work, helping formulate the last successful Senior Review proposal. His current research interests include plasma correlation scales in the solar wind.

The electron spectrometer that is being proposed for the DSCOVR spacecraft.

Judit Pap

Judit Pap has been working on solar-terrestrial physics. She conducts research on solar total and spectral irradiance variations, their relation to solar magnetic activity, modeling, and the role of solar variability in climate change. She has been working with images from the National Solar Observatory (magnetograms), Michelson Doppler Imager (MDI) magnetograms and quasi-continuum images, and is conducting collaborative research on these data with Harrison Jones from NSO/Kitt Peak under the NASA grant entitled "Bayesian Classification of Multidimensional NSO Magnetograms for Comparison with Solar Irradiance and Study of the Solar Cycle." She also collaborates with Luca Bertello of UCLA under the NASA grant entitled "The Fine Structure of Active Regions and Weak Magnetic Fields from MDI Images" (PI Judit Pap, Co-I Luca Bertello), and with Garry Chapman and Dora Preminger from the San Fernando Observatory at California State University.

The goals of her research is to identify the major features causing irradiance variations, to estimate their contribution to irradiance variations and to develop long-term models for climate studies. She is working with the SOHO/VIRGO[*], ACRIM[†], and SORCE/TIM[‡] total irradiance data, and the VIRGO/Sun Photometers and SORCE/Spectral Irradiance Monitor (SIM) spectral data. She is a Co-I on SOHO/VIRGO and MDI, a Co-I on the French PICARD experiments, and was selected as a Co-I on SDO's Helioseismic and Magnetic Imager (HMI). Her interest also covers variations in X-ray and EUV and the relation of these variations to solar magnetic field variations. In addition to her research, she works together with engineers at the Jet Propulsion Laboratory on the fabrication of a small radiometer and has been involved in mission planning for that instrument. In addition to her research, she is leading the Solar Variability group within the Scientific Committee of Solar Terrestrial Physics (SCOSTEP) Climate and Weather of the Sun–Earth System (CAWSES) international research organization; she is one of the six US disciplinary scientists in SCOSTEP. She organized the scientific program of the last CAWSES meeting at Bozeman, Montana as Chair of the Scientific Organizing Committee and she is the lead editor of a book on solar-terrestrial physics to be published by Cambridge University Press.

---

[*] VIRGO: Variability of Solar Irradiance and Gravity Oscillations
[†] ACRIM: Active Cavity Radiometer Irradiance Monitor
[‡] TIM: Total Irradiance Monitor

Etienne Pariat

Dr. Pariat is a solar physicist and a Post-Doctoral Research Assistant of the College of Science of George Mason University. He has been working as a contractor in the Space Weather Laboratory (Code 674) in collaboration with Dr. S. Antiochos since February 2008. He wrote two refereed papers on STEREO observations of coronal jets and state-of-the-art tri-dimensional numerical simulations of their generation of solar coronal. This work was presented at several international conferences and was presented as the topic of a press conference during the spring joint meeting of the AGU and Solar Physics Division. He also gave two invited talks on the generation of intense currents in the framework of magnetic reconnection. He has been associated with a review on magnetic helicity. Finally, Dr. Pariat has been representing the Heliophysics division at the Smithsonian Folklife Festival.

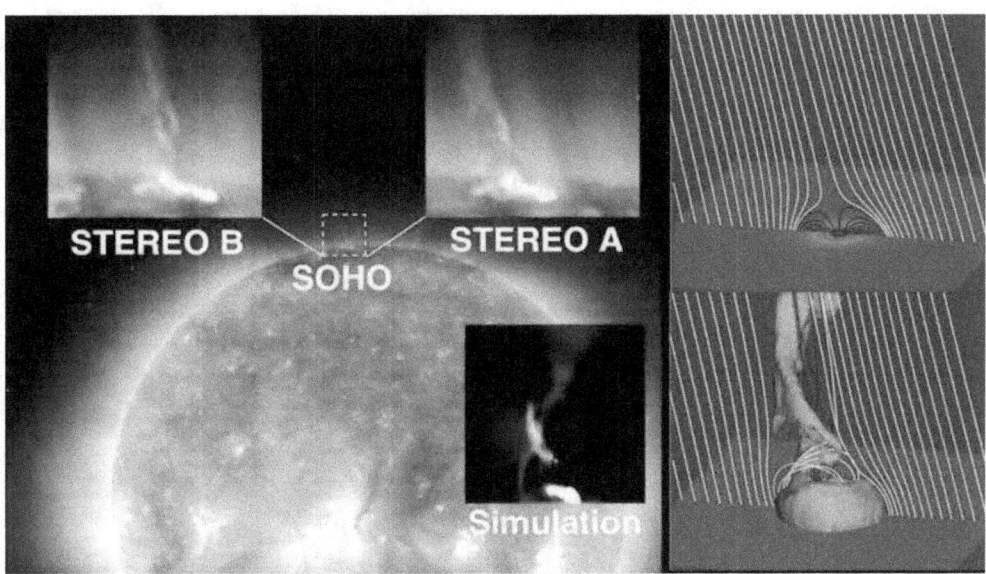

STEREO's two views (upper left and right) of a jet that erupted from the Sun's outer atmosphere in June 2007 show a twisted structure (orange), an indication that tangled magnetic fields propelled the jet. White indicates bright structures at the jet's base. the middle image shows the jet as seen by the single perspective of the SOHO spacecraft. The lower right image shows the output of the 3-D numerical simulation that reproduces the helical geometry of the observed solar jet.

Dean Pesnell

The year 2008 was a busy one for research and outreach. Michael Kirk finished the analysis of the area of the polar coronal holes and wrote a paper that has been accepted by *Solar Physics*. The prediction of Solar Cycle 24 continued to occupy a prominent spot in my research. A paper analyzing the known predictions appeared in *Solar Physics*, while talks describing how to predict solar activity were given in several venues. Both efforts are related to the long-term manifestations of the solar dynamo in observations. Peter Williams came to Goddard as a post doc to study observable changes in the convective velocity spectrum. Dr. Pesnell was awarded the Robert H. Goddard Award for Exceptional Achievement for Outreach for the Family Science Night Team and the NASA, Group Achievement Award for the TIMED Team.

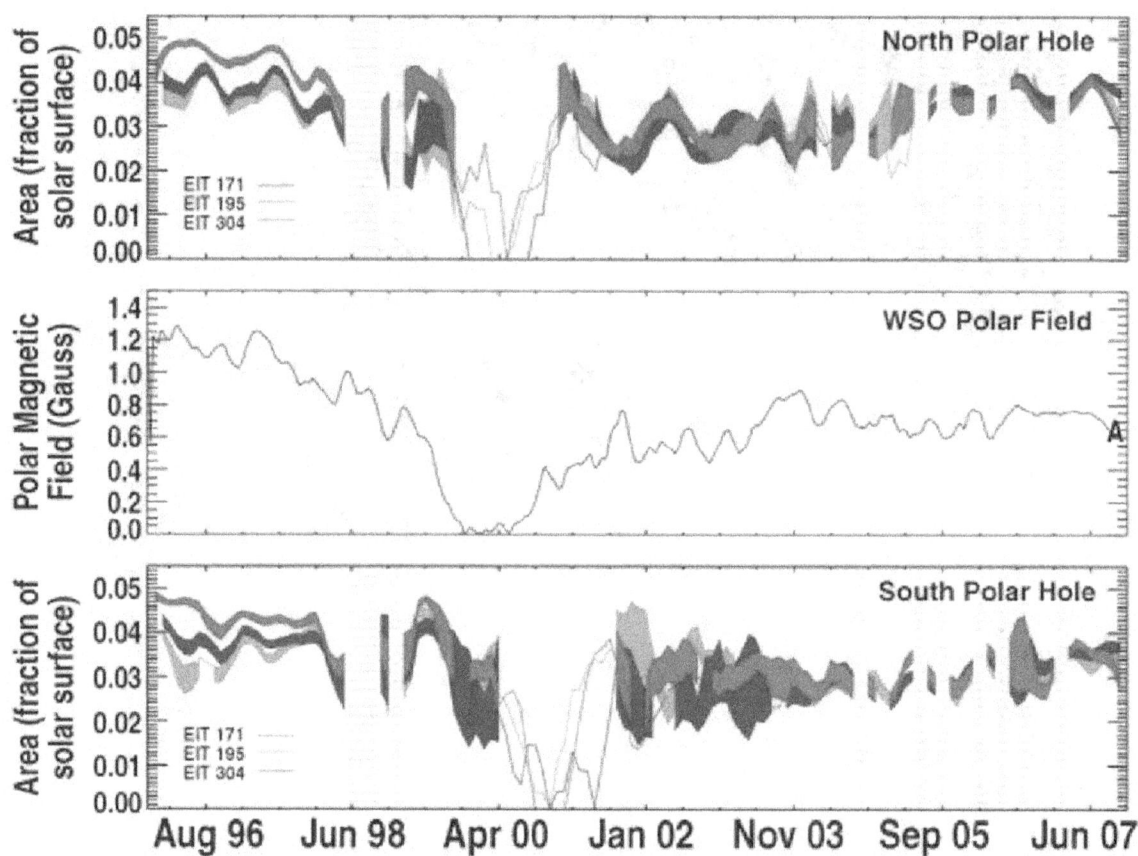

The time evolution of the polar coronal hole area for Solar Cycle 23.

Robert Pfaff

Dr. Robert F. Pfaff is a space scientist in the Space Weather Laboratory (Code 674). During FY08, his main research activities involved the successful launch and operation of the Vector Electric Field Investigation (VEFI) instrument suite on the C/NOFS satellite that was launched in April 2008 into a low latitude (13° inclination) orbit with perigee and apogee of 401 and 867 km, respectively. Dr. Pfaff is the PI of VEFI that returns vector DC and AC electric fields, magnetic fields, relative plasma density, and optical lightning flash count rates. The main objective of the instrument and the C/NOFS mission is to investigate the nature and cause(s) of low latitude ionospheric irregularities, as well as to provide data to be used to predict their occurrence.

As part of his LWS Targeted Research and Technology (TR&T) grant, Dr. Pfaff demonstrated how low and mid-latitude irregularities are associated with magnetic storms on a global basis (i.e., simultaneously at low and mid latitudes), as illustrated in the attached figure and discussed in Pfaff et al., 2008.

In FY08, Dr. Pfaff prepared two comprehensive, vector electric field experiments, for which he was the instrument PI, which were successfully launched into the active cusp on two simultaneous sounding rockets from Andoya, Norway in December 2007 (Dr. C. Kletzing, Univ. of Iowa, was the mission PI). The electric field experiments were both highly successful, returning DC and AC electric fields (including measurements of high frequency [HF] Langmuir waves) and absolute plasma density data.

Electric field and impedance probe experiments were also prepared in FY08 in Goddard's electric field laboratory for two auroral zone sounding rocket experiments with simultaneous launches planned for January 2009, from Poker Flat, Alaska.

Dr. Pfaff serves as the Project Scientist on the FAST mission, the CINDI investigation on C/NOFS, and the NASA Sounding Rocket Program. He was awarded the Robert H. Goddard Award for Exceptional Achievement in Science.

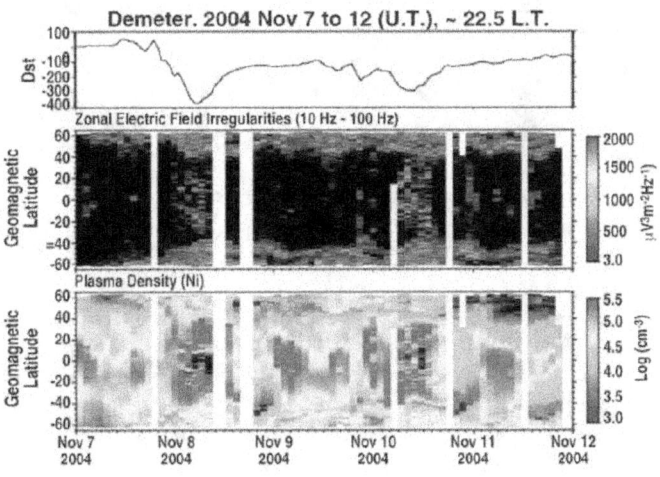

DEMETER satellite nighttime data (near 22:30 LT) for a five-day period from November 7–12, 2004 during a double peaked magnetic storm. The top panel shows the Dst, whereas the middle and lower panels show the integrated electric field irregularity power from 10–100 Hz and the plasma number density along each orbit. The satellite data have been converted to color scales and mapped to bins of magnetic latitude between –65 and +65 degrees.

Antti Pulkkinen

Dr. Pulkkinen, of UMBC, has been working on integrating and using space physics models hosted at the CCMC for the purposes of forecasting the ground effects of space weather. Over the past year, he has written several papers on theoretical and practical aspects of the ground-based space weather modeling. One of the more significant aspects of his latest work involves the use of heliospheric MHD models in long lead-time forecasting of geomagnetically induced currents (GICs) in North America's high-voltage power transmission system. Dr. Pulkkinen has presented his work in a number of scientific meetings in both regular and invited talks. He also led the development of CCMC's Google Earth experimental public space weather service and space weather show for Science on a Sphere.

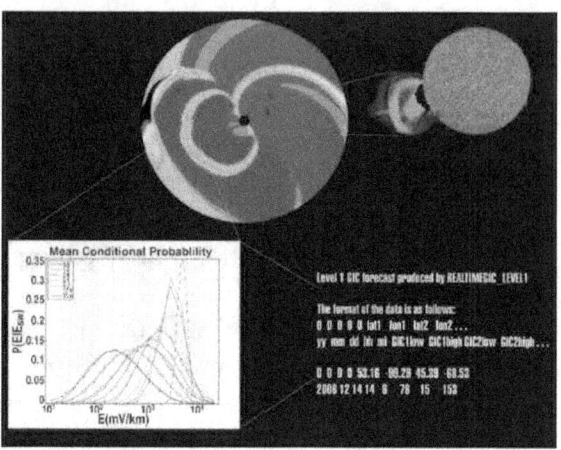

Design for the 1–2 day lead-time experimental GIC forecasting system. The system has been running in real-time at CCMC since February 2008. (Brosius, Rabin, and Thomas, 2008, Astrophys. J.)

Douglas Rabin

Dr. Rabin's research is focused on the dynamics of the solar transition region and corona. He is the PI of the Extreme Ultraviolet Normal Incidence Spectrograph (EUNIS) sounding rocket instrument, which had its second successful flight in November 2007. The unprecedented sensitivity of EUNIS allows it to probe intensity and velocity variations in solar plasmas on timescales as short as 10 s. The figure shows a result from the first EUNIS flight in 2006. The EUNIS Science Team won the 2008 Robert H. Goddard Award for Exceptional Achievement for Science. Dr. Rabin is the Deputy Project Scientist for SORCE. He helped organize and delivered the workshop summary at the meeting "Solar Variability, Earth's Climate, and the Space Environment," held in June 2008, co-sponsored by the NASA LLWS and International Heliophysical Year (IHY) programs.

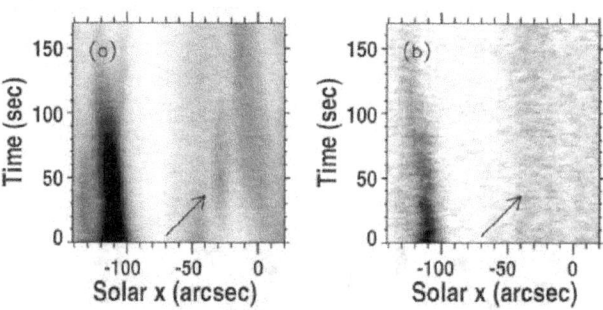

Space-time diagrams showing a short-lived transient event (left, negative intensity scale) in He II 30.4 nm observed by the EUNIS instrument capturing spectral images every 2.4 s. The transient is not visible in the hotter line of Mg IX 36.8 nm (right).

Lutz Rastaetter

Lutz Rastaetter is working with Michael Hesse and Maria Kuznetsova in the Space Weather Laboratory (Code 674) on numerical modeling and visualization at the Community Coordinated Modeling Center (CCMC).

He worked with numerous researchers to put observations (e.g., satellite measurements) into the context of multi-dimensional numeric models. Ongoing collaborations include the computation of synthetic Tomson scatter images from a proposed lunar vantage point using a magnetospheric simulation by the Solar Wind Modeling Framework (SWMF) performed at CCMC.

Dr. Rastaetter also performs verification and validation studies of (primarily magnetospheric) models resident at CCMC. Currently, he is working on modeled responses in cross polar-cap potentials after specific solar wind changes and on data analysis of CCMC simulation runs for the 2008 Geospace Environment Modeling (GEM) challenge.

Nelson Leslie Reginald

Dr. Reginald joined the Solar Physics Division (SPD) in January 1998 to conduct his doctoral dissertation work under the guidance of Dr. Davila on a NASA fellowship awarded through the University of Delaware. The dissertation work entailed designing and building the Multi-Aperture Coronal Spectrograph (MACS) to simultaneously measure at multiple lines of sight in the low solar corona the electron temperatures and its bulk flow speeds through the measurements of the K-coronal spectra and the theoretical modeling of the observed shapes of the K-coronal spectra to derive the temperatures and bulk flow speeds. The first measurements using MACS were conducted in conjunction with the total solar eclipse of 1999 August 11 in Elazig, Turkey. Upon graduation in January 2001, Dr. Reginald has continued his work in the SPD at Goddard in improving the MACS instrument and designing its companion instrument—the Imaging Spectrograph of Coronal electrons (IMACS)—to obtain maps of electron temperatures and flow speeds in the low solar corona. These instruments have been successfully used in conjunction with the total solar eclipses of 2001 June 21 in Zambia, 2006 March 29 in Libya, and 2008 August 1 in China. He is currently working on conducting this same experiment using the ground based Solar-C coronagraph located at the Mees Solar Observatory in Haleakala, Hawaii.

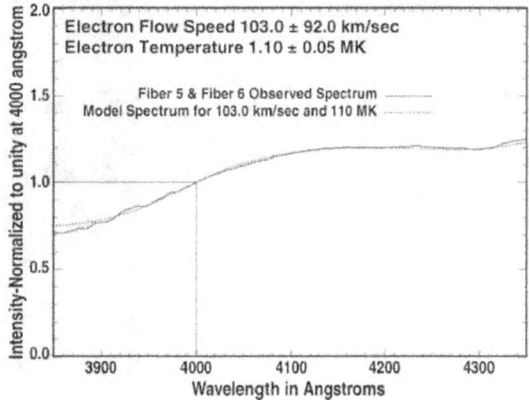

Matching the shape of an observed K-coronal spectrum with a theoretical model

Michael Reiner

Dr. Reiner (of CUA) has been working on the observations and analyses of solar radio emissions from a variety of space missions since coming to Goddard in 1984. The past year has been primarily devoted to the calibration of the radio instruments on the STEREO spacecraft and the analysis of solar radio emissions observed simultaneously from the STEREO and Wind spacecraft. A primary objective is to use these three-spacecraft stereoscopic observations to remotely observe, locate, and track solar radio sources through the interplanetary medium, to reveal the underlying magnetic field topology and to deduce the intrinsic radiation characteristics of these radio sources. He has recently submitted a paper that uses the direction-finding capabilities on STEREO and Wind to locate remote solar type III radio bursts by three-spacecraft triangulation and to provide the first three-point observations and analyses of the characteristic beaming patterns of these radio sources.

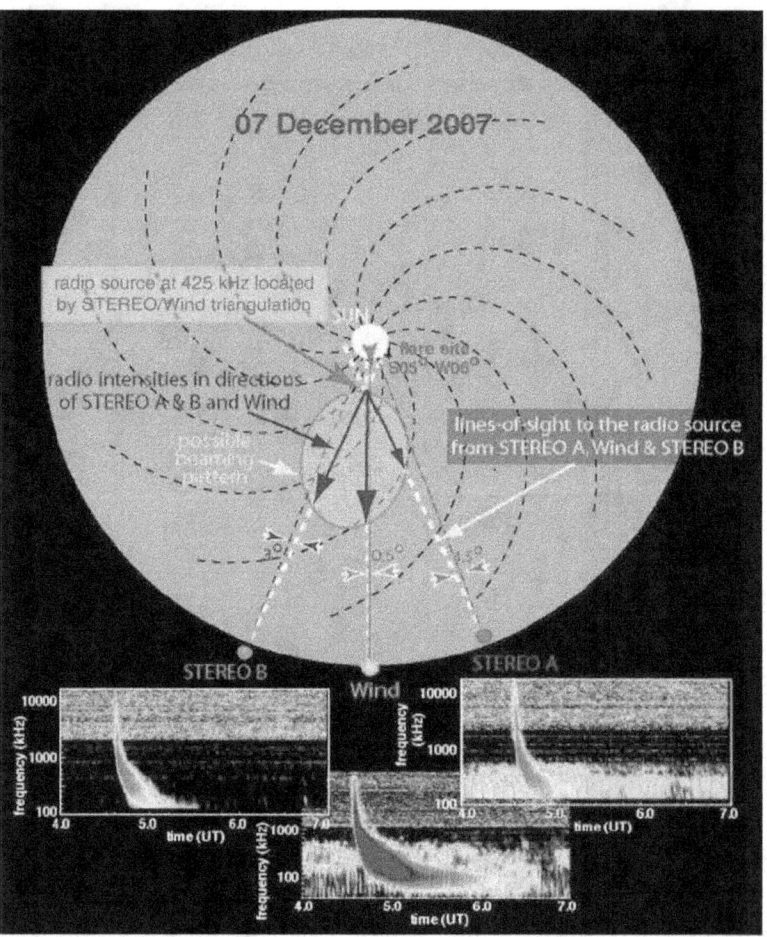

The triangulated source location for a type III solar radio burst (shown in the colored dynamic spectra below) that was simultaneously observed by the STEREO and Wind spacecraft on December 7, 2007 when the STEREO separation was 45°. These stereoscopic observations provided the first three-point measurement of the beaming pattern of the radio source at 425 kHz, indicating that the beaming pattern was orientated along the tangent to the Parker spiral magnetic field line

D. Aaron Roberts

D.A. Roberts spent much of his time working at NASA HQ on the Heliophysics Data Environment as the major author of the Heliophysics Data Policy, and the Program Scientist for the new Virtual Observatory Program. He worked with HQ and the Heliophysics community to set up the Heliophysics Data and Model Consortium, which will be on a par with the Space Physics Data Facility and the Solar Physics Data Facility and will oversee the work of the HP VOs, Resident Archives, and related work that is intended to provide unified short- and long-term access to heliophysics data and models.

He has continued his research activities, presenting results at AGU meetings on various aspects of turbulence in the heliosphere, including spectral exponents and the field variance directions. In one significant publication, he and his co-workers showed that while the spectrum of the magnetic field in the solar wind has interesting large scale features, these features are not consistent with the frequently cited "Fisk" model that includes the effects of differential rotation and dipole tilt of the Sun. It is still possible that the effects the Fisk model predicts are obscured by other solar variability, but it may also be that the coupling between the photosphere and the coronal magnetic field is not as the model implies.

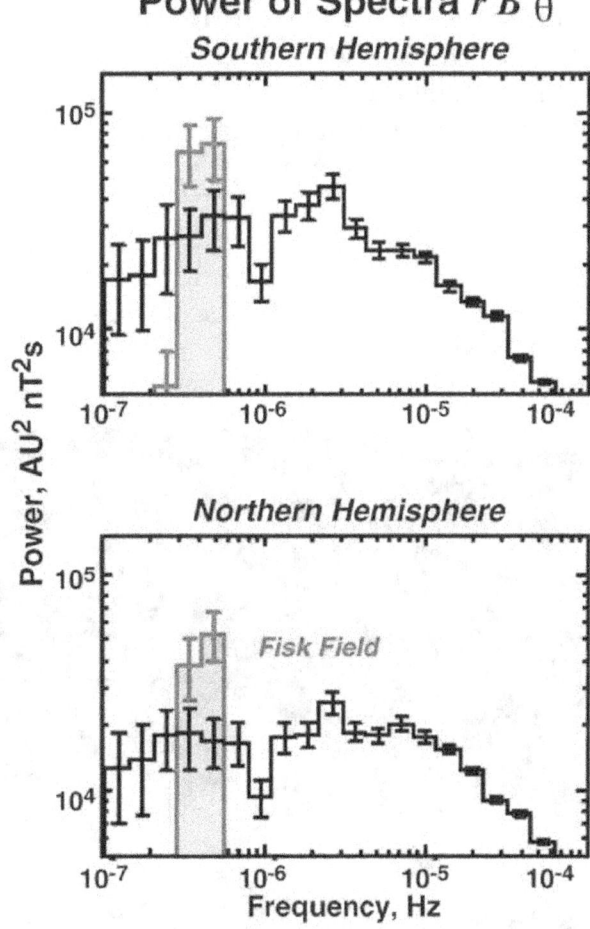

Douglas Rowland

Dr. Rowland is a space physicist in GSFC's Heliophysics Division since November 2003. His research continues to focus on topics relating to magnetosphere-ionosphere coupling and energetic particle acceleration processes. This past year, he delivered instrumentation for the TRICE and Auroral Current and Electrodynamics Structure (ACES) sounding rocket missions. The ACES mission will launch from Poker Flat, Alaska, in January 2009.

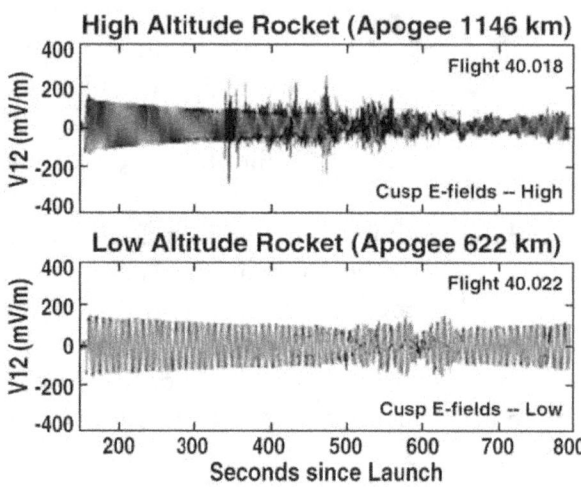

TRICE launched successfully into the dayside cusp, making simultaneous electric field and particle measurements on two separate rockets, one with an apogee of 622 km, and the other with an apogee of 1145 km. The goal of the TRICE mission was to determine if cusp reconnection signatures indicate spatially localized (called "patchy") or temporally localized ("bursty") reconnection, or both.

The TRICE sounding rocket DC electric fields are being analyzed to learn the implications for reconnection signatures in the cusp.

Dr. Rowland was also named as the PI for the second NSF CubeSat mission, "Firefly," which will determine the source of Terrestrial Gamma-ray Flashes (TGFs), and will determine the extent to which TGFs are associated with energetic electrons that can escape the atmosphere to populate the inner radiation belt.

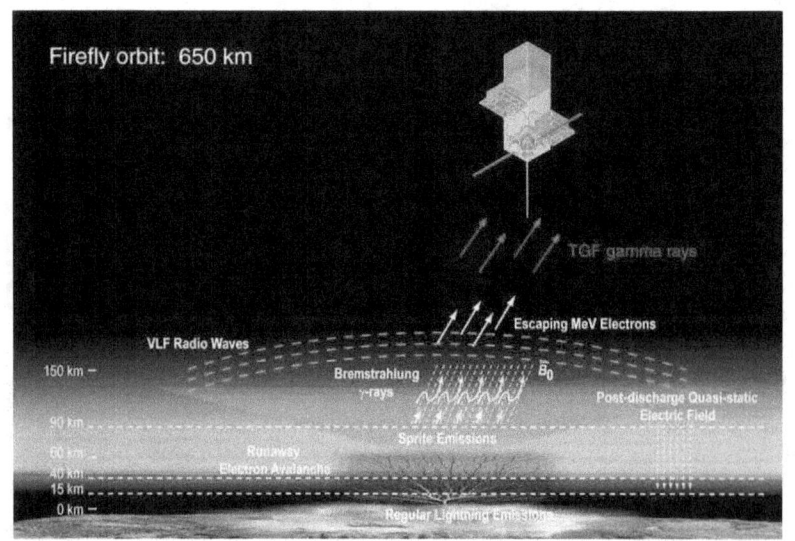

Firefly, scheduled for launch in 2010 / 2011, will determine if TGFs are caused by lightning, will measure MeV electrons associated with TGFs, and will push the detection envelope for TGFs to an order of magnitude weaker than previously observed.

Julia Saba

Dr. Saba, a solar physicist, is a long-term member of the Goddard solar group and a senior staff physicist of the Lockheed Martin Advanced Technology Center in Palo Alto, California. Dr. Saba helps with science planning and operations for the Michelson Doppler Imager (MDI) on SOHO. On the science side, she has been working with Dr. Keith Strong on potential new clues to the solar activity cycle, including the sudden onset to both cycles 22 and 23 and a recurring pattern in the magnetic evolution at the photosphere. Comparison of the strong-field magnetic flux at mid-latitudes to that near the equator appears to yield a useful forecast tool for upcoming solar activity. Dr. Saba is also interested in active region and flare dynamics and in coronal composition, and has been interacting with RHESSI colleagues and their visitors on these topics.

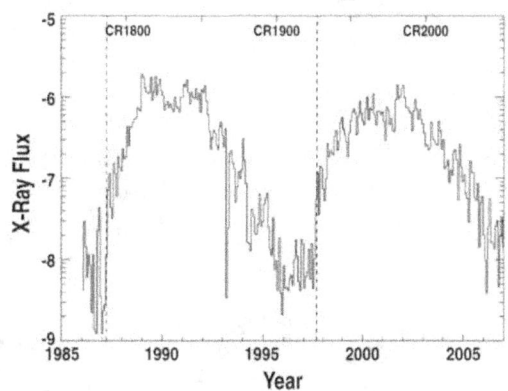

The soft X-ray flux for cycles 22 and 23. The sharp rises in 1987 and 1997 mark the onset timing of solar cycles 22 and 23 (Adv. Space Res.,accepted 2008).

Fouad Sahraoui

Dr. Sahraoui is a space plasma physicist who joined Code 673 in October 2007 as a visitor from the Centre National de la Recherche Scientifique (CNRS/CETP[*], France). He has been doing research at GSFC to understand plasma turbulence in the magnetosphere and in the solar wind—both theoretically and observationally. Working with Dr. Goldstein, a new method of investigation coherent structures and intermittency has been developed. Its application to the Cluster spacecraft data has yielded evidence of new properties of low frequency compressible magnetospheric turbulence. Dr Sahraoui is currently investigating small-scale solar wind turbulence, where a new process of energy dissipation is revealed. Dr. Sahraoui gave several presentations and seminars (e.g., at GSFC and the Space Sciences Laboratory–Berkeley) and two invited talks at the Institute of Mathematics-University of Warwick (United Kingdom) on "Structures and Waves in Anisotropic Turbulence" (October 2008), and at the 15th Cluster Workshop (Tenerife, Spain, March 2008).

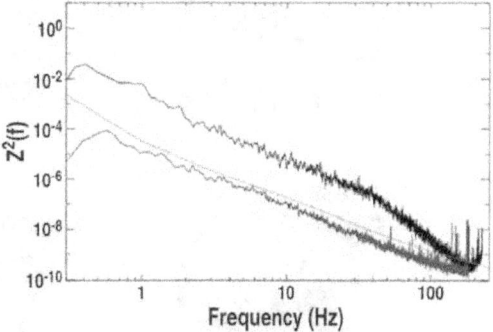

Spectra of small scale magnetic turbulence (from the Cluster STAFF-SC data). Evidence of a break point of energy cascade at the electron gyroscale scale.

---

[*] CETP: Centre de Recherches en Physique de l'Environment Terrestre et Planetaire

Menelaos Sarantos

Funded by an NPP Fellowship, Dr. Sarantos is investigating the plasma interaction with the collisionless atmospheres of Mercury and the Moon. He has applied models of Mercury's neutral sodium cloud to study the production and subsequent redistribution of pickup ions in Mercury's magnetosphere, helping interpret data that were obtained by the MESSENGER[*] spacecraft during its first two flybys. Lunar research has focused on developing models for the re-supply of exospheric species by the lunar surface via various physical processes. Such tools may be applied to the planning of observations that are expected to be obtained by the proposed Lunar Exosphere and Dust Environment Explorer (LADEE).

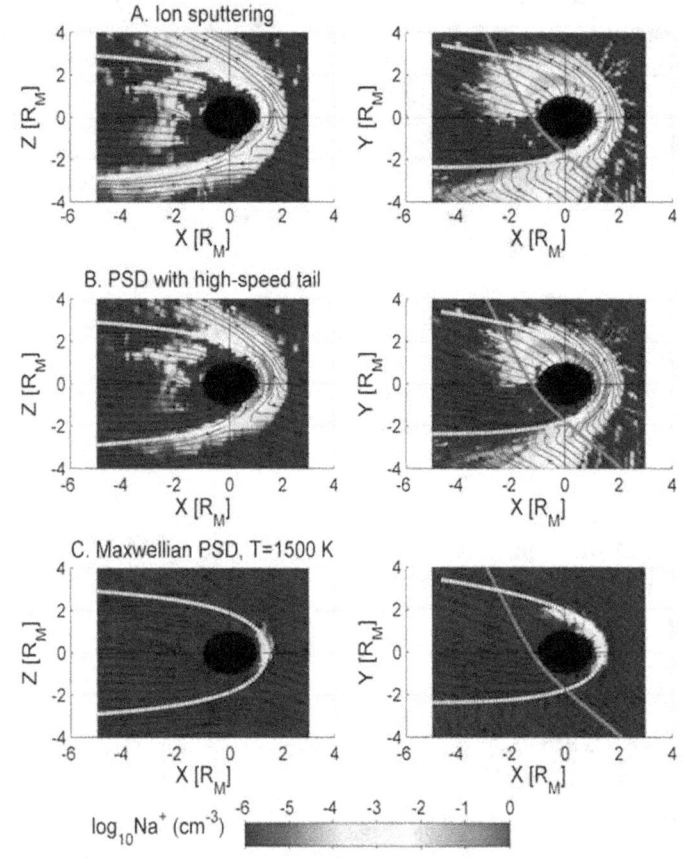

The modeled distribution of sodium pickup ions inside (and about) Mercury's magnetosphere at the time of MESSENGER's first Mercury flyby. The ions are initialized according to various neutral surface release processes (sputtering and/or photodesorption), and their trajectories are followed using single-particle tracing and fields given by a magnetohydrodynamic simulation.

---

[*] MESSENGER: Mercury Surface, Space Environment, Geochemistry and Ranging

Chris St. Cyr

Dr. St. Cyr has been a scientist in GSFC's Heliophysics Science Division since December 2002. He is a Co-I on STEREO's SECCHI and SWAVES instruments, and in the past he was an operations scientist for the Solar Maximum Mission and for SOHO. His research interests include the initiation and propagation of solar coronal mass ejections; testing new instrument techniques at total solar eclipses; and the quantification of economic impacts of space weather in electric power grids. During 2008, he worked with the Office of the Chief Engineer at NASA HQ to inventory space weather requirements across the Agency; and he has been a Co-Chair for the triennial Community Roadmap strategic planning activity for the Heliophysics Division.

Richard Schwartz

During the past year, Dr. Schwartz concentrated on improving critical elements of the RHESSI analysis software, as well as fixing any problems in the base data acquisition software. Improvements were made in understanding and deconvolving the effect of pulse pileup. In a related effort, Dr. Schwartz has tried to understand the discrepancies in RHESSI photon spectra on either side of attenuator crossings. He is preparing several improvements in the RHESSI software to increase the speed of image analysis for the CLEAN and PIXONS algorithms and the accuracy of spectral accumulations.

His recent work involves comparing the RHESSI spectrum below 10 keV with that from the MESSENGER X-ray spectrometer and preparation for data from the SPHINX spectrometer included in the upcoming CORONAS payload.

Dave Sibeck

As Project Scientist, Dr. Sibeck led efforts to publicize THEMIS science discoveries in 2008 on topics including substorms, flux transfer events, and hot flow anomalies. He set up the outline for, compiled, and edited the THEMIS senior review proposal, writing major portions of that document. He was engaged in THEMIS E/PO activities at the Fall AGU, the Smithsonian, Baltimore Science Museum, and GSFC throughout the year. He was a frequent speaker on CBC radio stations and participated in television broadcasts throughout the United States. Dr. Sibeck led or participated in 8 proposals to NASA/HQ, and was either the lead or co-author on 13 publications. D. G. Sibeck participated in a complete review of all NSF Space Physics activities, providing detailed recommendations on how these might be improved. He worked closely with Mike Collier on proposals for a mission to image Earth's magnetosheath.

Dr. Sibeck was a major organizer of the 2008 Huntsville Meeting, inviting about 25% of the speakers and defining the session topics. He continued to serve as campaign coordinator for all bow shock, magnetopause, and cusp sessions at the NSF's GEM Winter and Summer meetings. By year's end, he was actively planning an International Living With a Star (ILWS) meeting to be held in Guaruja, Brazil during October 2009. Dr. Sibeck continued to serve as the corresponding editor for Space Physics at EOS and as the editor for all heliospheric papers submitted to *Advances in Space Physics* throughout the year. He became the Mission Scientist for RBSP somewhere between October 2008 and January 2009. In this position, he continued to advise both NASA and the Johns Hopkins University's Applied Physics Laboratory on mission requirements. At the end of the period covered, he attended the RBSP Preliminary Design Review (PDR) and all but one instrument PDR.

John B. Sigwarth

Dr. Sigwarth is leading an effort to develop a new generation camera for ionospheric, thermospheric, and auroral remote sensing. This new camera, the Thermospheric Temperature Imager (TTI) is designed with the capability to remotely sense the temperature of the thermosphere, as well as the thermospheric composition and structure. In addition, the TTI camera is capable of determining the line-of-sight velocities of the circulating constituents that give rise to auroral emissions. Dr. Sigwarth has briefed this new design to the US Navy and Department of Defense (DoD) Space Experiment Review Boards (SERB) and the TTI camera has been placed on the SERB list for possible flight opportunities arranged by the DoD Space Test Program (STP). As a consequence of inclusion on the SERB list, the TTI camera will be flown in December 2009 on the Fast, Affordable, Science and Technology Satellite (FASTSAT) mission of the STP. In his recent research, Dr. Sigwarth has co-authored papers on the aurora and magnetospheric-ionospheric coupling, and presented conference papers on the relation of the thermosphere temperature and composition using the TIMED Global Ultraviolet Imager (GUVI) and the Polar Visible Imaging System (VIS) to obtain a 3-dimensional understanding of the processes in the thermosphere. As part of his duties at GSFC, Dr. Sigwarth leads the efforts for technology development in support of science in the Heliophysics Science Division.

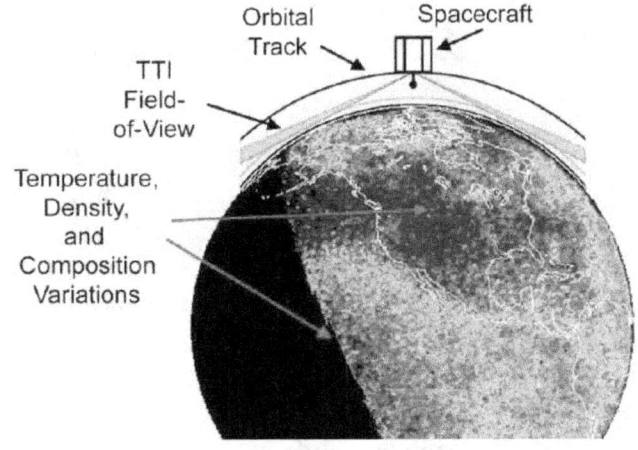

Field of view for the TTI from the FASTSAT orbit. Temperature changes in the thermosphere drive composition changes in the upper atmospheric layer.

Edward C. Sittler, Jr.

Dr. Sittler is an astrophysicist in Code 673, who continues to work in a broad spectrum of interests including heliospheric physics, planetary magnetospheres, and astrobiology. He is a Co-I on the Cassini Plasma Spectrometer (CAPS) team and the PI on three other projects: (1) a Cassini Data Analysis Program (CDAP)-funded effort to analyze CAPS plasma data within Saturn's inner magnetosphere and its moon Enceladus; (2) development of an LWS instrument development ion mass spectrometer (IMS), which is now a working prototype, to provide mass and charge state analysis of the solar wind ion composition, PI on an LWS

Ed Sittler visited the Lockheed Martin Vortek Lamp Facility developing high temperature plasma wave electric field antenna for possible SP+ application.

TR&T-funded effort to develop a semi-empirical MHD model of the solar corona and solar wind; and (3) a recently funded instrument development effort under the astrobiology instrument development (ASTID) program to build a 3-D ion-neutral mass spectrometer (INMS) for a mission to Europa, in collaboration with Code 699, headed by Dr. Paul Mahaffy. Dr. Sittler is also a member of the flagship Titan Saturn System Mission (TSSM) science study team co-chaired by Jonathan Lunine from the University of Arizona and is a Co-I on the funded ESA Tandem (Titan and Enceladus Mission) proposal headed by Dr. Athena Coustenis (Observatoire de Paris, Meudon, France).

Dr. Sittler has been very active in studying Saturn's magnetosphere and its interaction with its moon Titan. The plasma environment of Saturn's magnetosphere is dominated by its moon Enceladus, similar to the role Io plays with regard to Jupiter's magnetosphere, thus causing the magnetosphere to be distorted into a magnetodisk configuration, which then determines how Saturn's magnetosphere interacts with Titan. Titan has a very dense extended upper atmosphere of nitrogen and methane gas, which has a very complex hydrocarbon and nitrile chemistry due to the interaction of solar UV and Saturn's magnetosphere with Titan. Because of Enceladus, Saturn's magnetosphere bombards Titan's upper atmosphere with keV water group ions, which can then become trapped in Titan's aerosols, which then settle on its

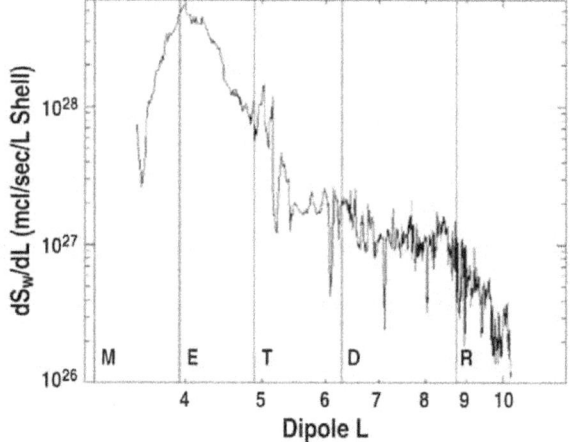

Neutral production rate per unit L shell plotted versus dipole L shell. Icy satellite L shells are indicated. We used the beutral cloud model in Johnson et al, (2006) and Burger et al. (2007). It shows large peak centered on the L shell of Enceladus wih total source $S_w \sim 2 \times 10^{28}$ mol/s.

surface with potential exobiology implications. Under the LWS TR&T program, a recent breakthrough provides an accurate description of the Sun's corona and solar wind in 2-D under solar minimum conditions. These results will be presented at the upcoming Fall AGU. With regard to the ASTID-funded effort, for which he also received mid-term IRAD funding, he is developing a design concept that can separate ions with mass resolution $M/\Delta M \sim 10{,}000$ (i.e., separate ions of similar mass such as N_2^+, CO^+, and $HCNH^+$). This design can separate isotopes and resolve species at parts-per-billion levels. Dr. Sittler has already submitted one patent request and more are planned. Recently, he is also developing a nadir-viewing concept that is planned to be proposed for SP+.

James Slavin

Dr. Slavin, the Director of HSD, is a magnetospheric physicist. Dr. Slavin's research continues to focus on planetary magnetospheres with an even balance between Earth and the other terrestrial planets. Dr. Slavin proposed the ST5 mission to NASA in 1999 and led it through launch as the Project Scientist. This past year, he used ST5's measurements as a magnetic field gradiometer to separate—for the first time—temporal and spatial variations in field-aligned currents in low Earth orbit. He is also the lead Co-I for magnetospheric magnetic fields on the MESSENGER mission to Mercury. Initial results from MESSENGER's first fly-by on 14 January 2008, revealed a complex interaction between this planet's miniature magnetosphere and its tenuous atmosphere with newly created planetary pickup ions extending even upstream of the bow shock and magnetopause.

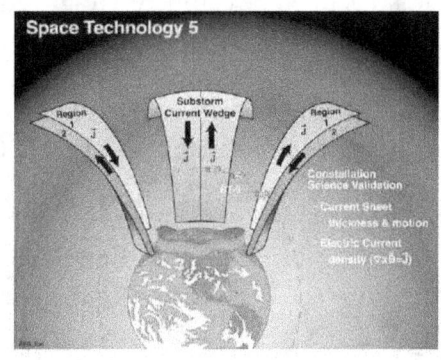

Space Technology 5 mission yields first gradiometric measurements of field-aligned currents and crustal magnetic fields.

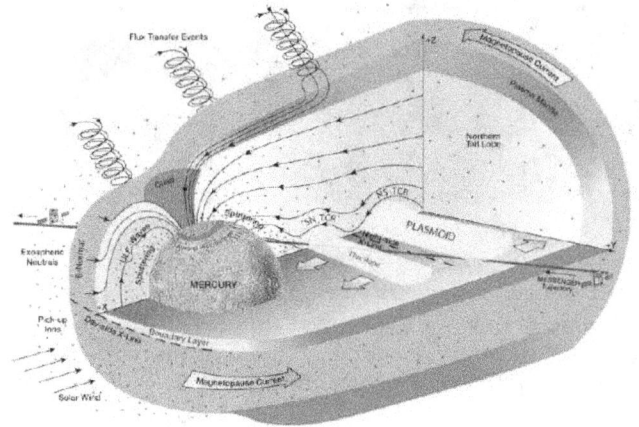

MESSENGER mission found that Mercury's miniature magnetosphere is immersed in a comet-like cloud of planetary ions.

Keith Strong

Dr. Strong is a solar physicist with SP Systems, who joined Code 670 in March 2008. He has been doing research at GSFC on understanding and predicting the solar cycle, producing several presentations and drafting three papers. Working with Dr. Saba, a new timing reference—the onset—for the solar activity cycle has been established. A way of tracing the evolution of the cycle and predicting the onset using various specific flux and flux distribution ratios has been developed. He gave an invited talk on the future of space weather research at the NOAA Space Weather Workshop in Boulder, Colorado. Dr. Strong has worked with the American Meteorological Society to organize a session on space weather at its annual meeting, and has worked to develop a position statement for the society. He has also helped support Code 670's E/PO and recruitment efforts.

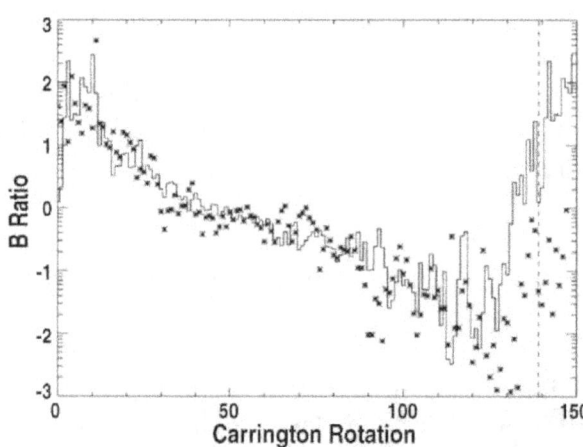

The soft X-ray flux for cycles 22 and 23. The sharp rises in 1987 and 1997 mark the onset timing (Adv. Space Res., accepted 2008).

Tim Stubbs

Dr. Stubbs, an assistant research scientist with UMBC/GEST, is currently pursuing research in surface charging and dust transport on the Moon and other airless bodies in the solar system. This has resulted in the development of the "dust fountain" model, lunar surface charging models, and a model for predicting lunar horizon glow. Dr Stubbs is a participating scientist with the Lunar Reconnaissance Orbiter (LRO) mission, and will be using the data to model lunar surface electric fields and search for evidence of exospheric dust. He has served on the Lunar Airborne Dust Toxicity Advisory Group, (LADTAG, 2005–present); the NASA Advisory Council (NAC) Heliophysics Lunar

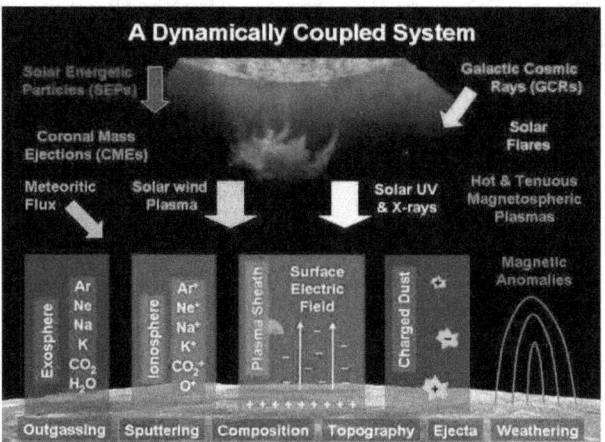

An overview of the dynamic lunar environment, showing how the well-known planetary processes (examples listed at the bottom) couple with external drivers (listed at the top) to produce the various interdependent components of the lunar environment.

Subpanel (2006–2008); and the NASA Engineering and Safety Center (NESC) Mechanical Systems Lunar Dust Assessment Team (2007–2008).

Adam Szabo

Dr. Szabo specializes in heliospheric and magnetospheric shocks and discontinuities. His work includes the extension of the MHD "Rankine-Hugoniot" shock jump condition fitting technique. In collaboration with A. Koval, he has extended this method to analyze ion temperature anisotropies and the alpha particle content at the shocks. Using millisecond time resolution magnetic field observations made by the Wind spacecraft, he has studied the microstructure of interplanetary shocks as a function of upstream conditions. This work has significant ramifications for particle acceleration processes.

Dr. Szabo is the Project Scientist for the Wind spacecraft and the PI for the magnetic field investigation on the same satellite. He is also the PI for the Virtual Heliospheric Observatory (VHO) and served as the Study Scientist for the LWS Sentinels mission.

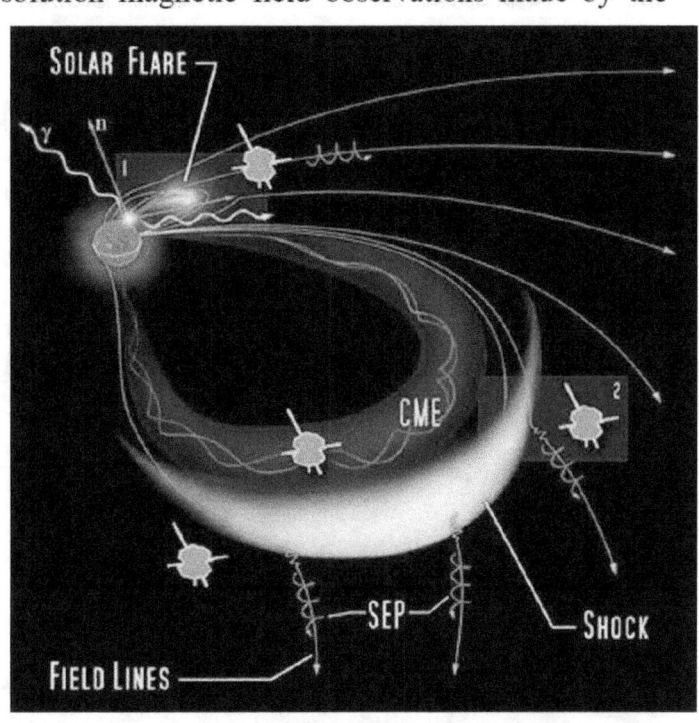

One possible orbital configuration of the four Sentinels spacecraft to investigate the topology of Coronal Mass Ejections and the acceleration of particles at the CME drive shocks and from solar flares (from the Sentinels STDT Report).

Roger Thomas

Dr. Thomas is an internationally recognized expert on the design and scientific use of extreme ultraviolet spectrographs. He created the optical design of the EUV Imaging Spectrometer (EIS) on the Hinode mission, as well the designs of seven sounding rocket instruments for which he is a Co-I: EUNIS, Solar EUV Research Telescope and Spectrograph (SERTS), Multi-Order Solar EUV Spectrograph (MOSES), Solar Ultraviolet Normal Magnetograph Investigation (SUMI), Rapid Acquisition Imaging Spectrograph Experiment (RAISE), Very High Angular Resolution Imaging Spectrometer (VERIS), and UV Spectro-Coronagraph (UVSC). He is also a Co-I on Hinode/EIS.

Dr. Thomas leads the radiometric calibration of the EUNIS instrument, a key aspect of its scientific value and utility for underflight calibration of orbital instruments. Calibrations are carried out in collaboration with the Rutherford Appleton Laboratory (United Kingdom) and Physikalische Technische Bundesanstalt (Germany). The figure shows the calibration of the long wavelength channel on EUNIS-06, for which the internal precision was reduced below 10% for the first time.

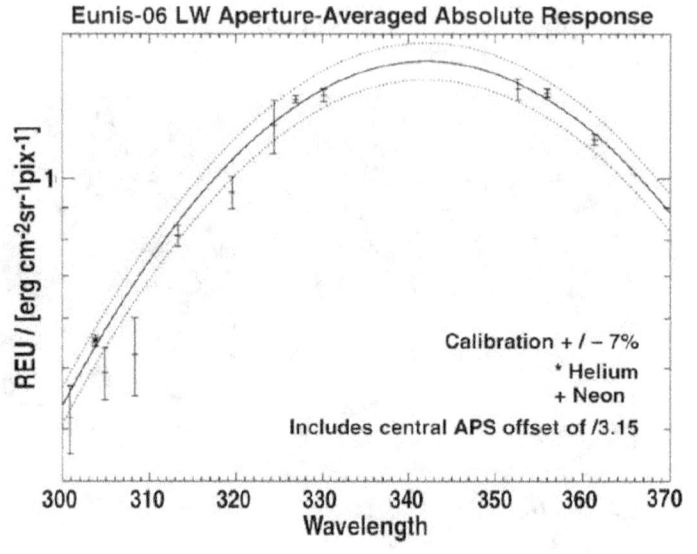

Absolute radiometric sensitivity of EUNIS-06 as a function of wavelength.

Barbara Thompson

Dr. Thompson, a solar physicist, joined Code 671 in August 1998. Her research has focused on the structure of, and phenomena associated with, coronal mass ejections. She served as the SDO Project Scientist from 2001–2004, and currently serves as the Deputy Project Scientist. She also serves as the Director of Operations for the International Heliophysical Year, and is the Science Corresponding Secretary for the International Living with a Star Program.

William Thompson

Dr. Thompson, of Adnet Systems Inc., serves as the Chief Observer for the NASA Solar Terrestrial Relations Observatory (STEREO) mission. As such, he oversees the STEREO Science Center, which serves as the primary archive for the mission, and is the processing point for the STEREO space weather beacon data. As Chief Observer, he is also responsible for coordinating scientific activities between the STEREO instrument teams. Dr. Thompson is also a member of the team operating the COR1 telescope aboard STEREO, and is responsible for characterizing the instrumental calibration. An inflight calibration of the instrument was established using observations of the planet Jupiter. An important part of the calibration of a coronagraph is background subtraction, and significant progress was made during FY08 in determining the time-dependent instrumental background. Dr. Thompson also serves on the IAU working group on the Flexible Image Transport System (FITS) standard.

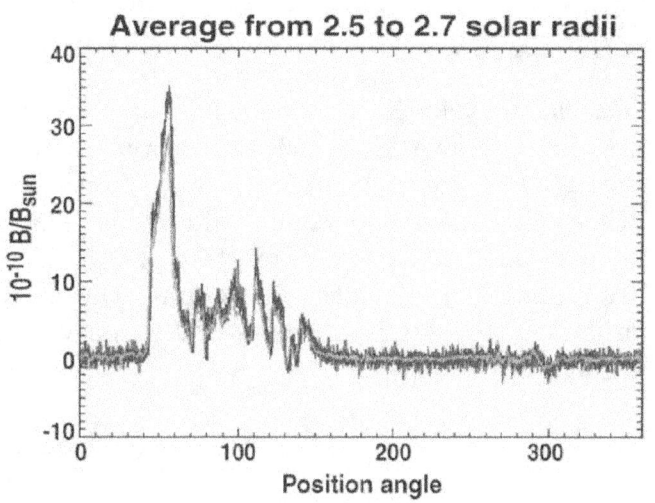

Comparison of COR1 Ahead, Behind and LASCO C2 intensities for a CME observed in January 2007.

Arcadi Usmanov

Dr. Usmanov, of the University of Delaware, has recently been working on developing a global three-dimensional solar wind model that describes properties of the large scale solar wind, interplanetary magnetic field, and turbulence throughout the heliosphere from the coronal base to 100 AU. The model is used to study the self-consistent interaction between the large-scale solar wind and smaller-scale turbulence, the role of the turbulence in the large scale dynamics and temperature distribution in the solar wind, and the effects of pickup protons in the physical processes of the outer heliosphere. He has also given scientific presentations at the AGU and ASTRONUM 2008 meetings.

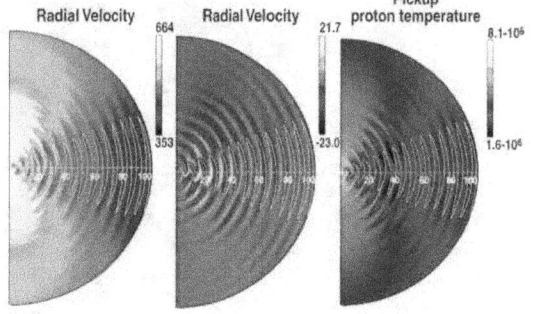

Meridional cut with contour lines of solar wind radial and meridional velocity, and pickup proton temperature for a dipole source field on the Sun, tilted by 30 degrees. The computational region extends from the coronal base to 100 AU. The white line represents the heliospheric current sheet.

Adolfo F.- Viñas

Dr. F.-Viñas is a space plasma physicist with the Heliophysics Science Division. He has also served as a research scientist at other GSFC's divisions since 1980. He is a Co-I on Wind's SWE and Cluster's PEACE instruments, and currently, he is a Co-I with the Fast Plasma Investigation of the Magnetospheric Multiscale Satellites (MMS) mission. Dr. Viñas' research interests include the study of kinetic and MHD processes, plasma instabilities, kinetic turbulence, and shocks and discontinuities in the solar corona, solar wind, and magnetosphere. He is currently involved in the simulation and testing of new analysis techniques for the Fast Plasma Investigation of the MMS mission. Dr. Viñas has developed a technique to describe shock and discontinuity of physical and geometrical properties, which has been implemented in a visualization and analysis tool named SDAT. Recently, he has also developed a new spherical harmonic spectral method for the calculation and modeling of particle velocity distribution functions (VDFs) and its moments and anisotropies for fast and high angular and energy resolution plasma instruments. An example of modeling Cluster core-halo, strahl, and reflected electron particles in the upstream solar wind region is included. He is currently supervising an NPP fellow.

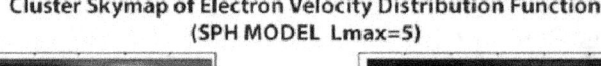
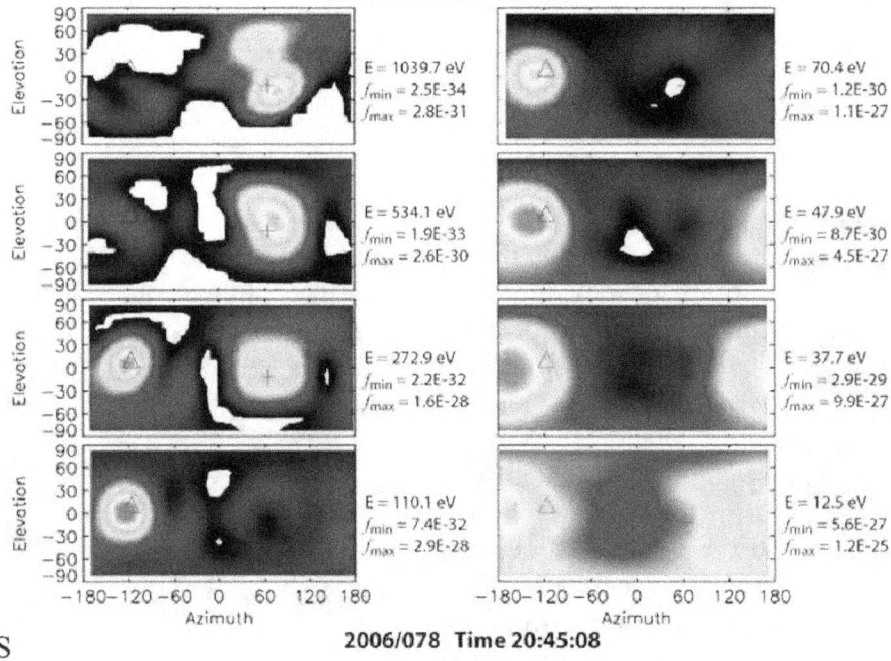

Tongjiang Wang

Dr. Wang is a solar physicist with SP Systems who joined Code 671 in May 2007. He has been doing research at GSFC on observations of oscillations and waves in coronal loops with the goal of understanding coronal heating and is exploring the application of coronal seismology to determine the poorly known physical parameters of coronal structures. Working with Prof. Ofman, the first evidence for transverse waves in coronal multi-threaded loops with cool plasma flowing along the threads was found. They also found the 5 min quasi-periodic oscillations in the transition region and corona near the footpoint of a coronal loop, revealing for the first time that the amplitude of oscillations decreases with increasing temperature. In addition, working with Drs. Brosius, Thomas, Rabin, and Davila, a technique was developed for deriving the absolute radiometric calibration of the EUNIS-06 short wavelength (SW) bandpass from these direct long wavelength (LW) results by means of density- and temperature-insensitive line intensity ratios. These results have been presented at SPD, AGU, and Hinode meetings.

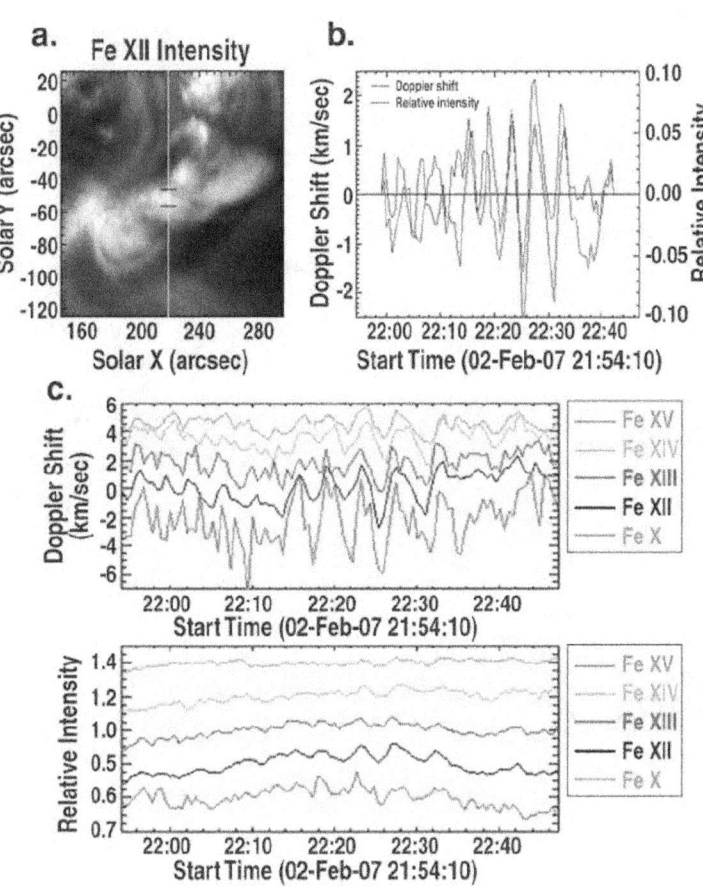

Propagating slow magnetoacoustic waves in a coronal loop observed by Hinode/EIS. (a) The intensity map in the Fe XII 195.12° line. The vertical line shows the position of the EIS 1° slit during the sit-and-stare observation. The short horizontal lines on the slit mark the range of rows where the oscillations are measured. b) Time profiles of averaged Doppler shift and relative intensity across the loop over 10 pixels. The in-phase relationship indicates the presence of an upwardly propagating acoustic wave. (c) Comparisons of time profiles of Doppler shift (upper panel) and relative intensity (bottom panel) in five coronal lines of Fe X-Fe XV, indicating a temperature dependence of the oscillation amplitude.

Yongli Wang

Dr. Wang is a space physicist with UMBC/GEST, who joined Code 674 in January 2006. He has been doing research at GSFC into understanding the dynamics of field-aligned currents (FACs) using multiple-spacecraft Space Technology 5 (ST5) observations and the characterization of flux transfer events (FTEs) using multiple-spacecraft Cluster observations. Working with Dr. Guan Le and Dr. James Slavin, he was a core member for the ST5 satellite magnetic field data calibration and public distribution. In 2008, Dr. Wang worked on FAC dynamics and FTE characterization and automatic identification. He has been either the author or coauthor of several papers that were published or submitted in this year. He has also been involved in two proposals, one of which was recently awarded with him as Co-I, and the other one is pending award, with him as PI. In addition, he has also been a very active reviewer for several leading journals, as well as for NASA and NSF, including serving on NSF's review panel.

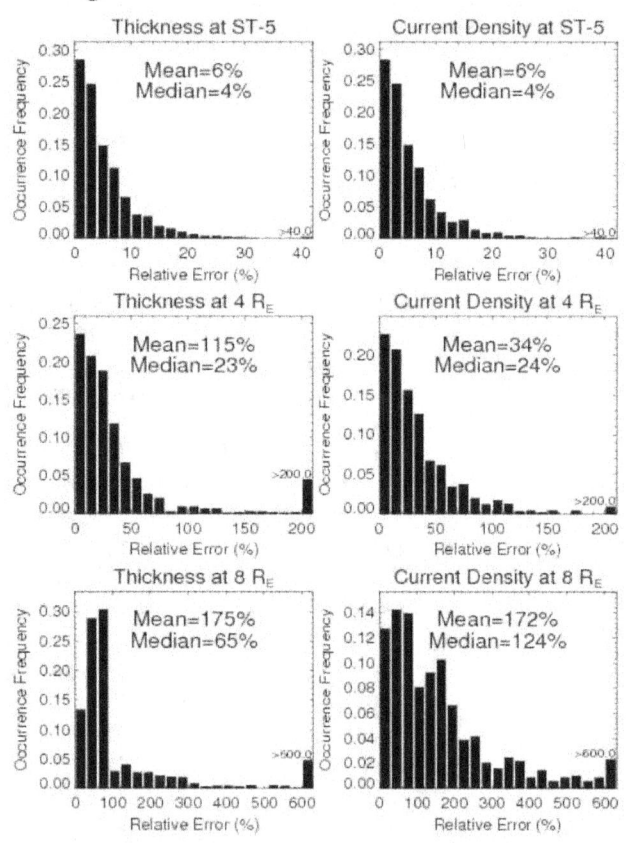

Relative errors of FAC thickness and current density, caused by FAC motion, at ST5 (300–450 km), 4 R_E and 8 R_E. This is the first systematic investigation of such dependence and it was found that the errors are negligible at ST5, quite large at mid-altitude, and very significant at high altitude.

Phillip Webb

Dr. Webb, of UMBC/GEST, has research interests primarily in the topside ionosphere and overlying plasmasphere. His ionospheric research centers on modeling studies of topside electron density profiles obtained from the International Satellites for Ionospheric Studies (ISIS) ionogram data set that encompasses a period of greater than two solar cycles (22 years). The focus of his plasmasphere research has been the development and refinement of an automated fitting technique to extract plasmaspheric electron densities from dynamic spectra obtained from the Radio Plasma Imager (RPI) on NASA's Imager for Magnetopause-to-Aurora Global Exploration (IMAGE) satellite, which collected data from 2000–2005. Dr. Webb has given a number presentations and posters at scientific meetings in the past year on these topics. Dr. Webb is a member of Dr. Laurie Leshin's GSFC Deputy Director's Council on Science (DDSC), which is tasked with providing the Deputy Director with science policy to improve GSFC competitive capabilities. Furthermore, he is the GEST Group Mentor for both the Heliophysics and Solar System Divisions.

Dr. Webb is a member of the board of the non-profit "The Inspire Project, Inc.," which seeks to prompt science education by teaching about lightning-generated very low frequency (VLF) electromagnetic waves and by selling an electronic kit that can be assembled to detect these natural emissions. In September, Dr. Webb began teaching a First Year Seminar (FYS) course at UMBC entitled "Chasing Lightning: Spherics, Tweeks, and Whistlers." The goal of this course is to introduce primarily non-science students to the wonders of VLF emissions from lightning and their detection from many thousands of kilometers away by use of the Inspire receiver kits.

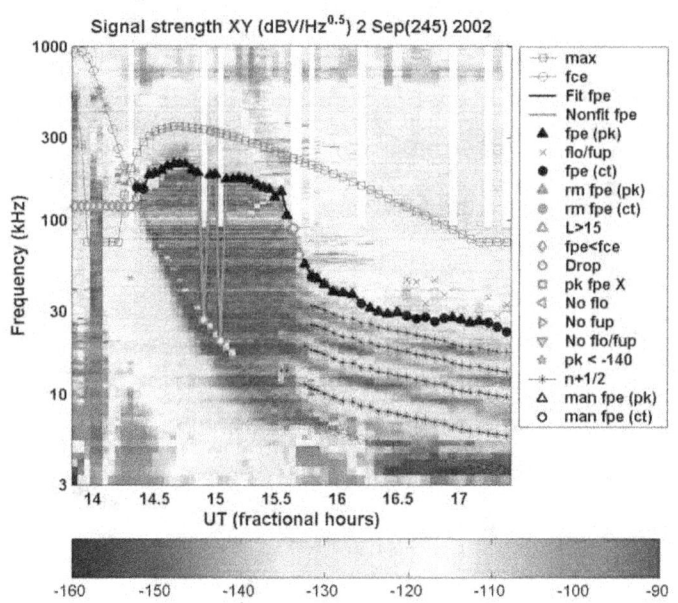

Example of the automated fitting of an IMAGE RPI plasmaspheric dynamic spectrum. The black solid triangles and circles denote the calculated electron plasma frequency obtained from successful fits to the upper-hybrid emission band and the lower edge of the continuum emission band, respectively.

Peter Williams

Dr. Williams has been analyzing Doppler velocity images from the Michelson Doppler Imager (MDI) aboard the Solar and Heliospheric Observatory (SOHO). These images can be reduced to extract surface manifestations of internal convection mechanisms, seen as either granule or supergranule cells. Of interest are various characteristics of these features such as their sizes, lifetimes, and advection properties. Current work involves the study of these, and similar, characteristics through an entire solar cycle to gain insight into how these surface features are influenced by changes in the global magnetic field of the Sun. Supergranules are significant in such studies as there is clear evidence of magnetic field interactions within, and at the boundaries of, such cells. A recent result has been the characterization of instrumental artifacts within the image that may be removed, along with the stronger velocity components of granules and supergranules, to aid the observational investigations of, hitherto elusive, giant cells.

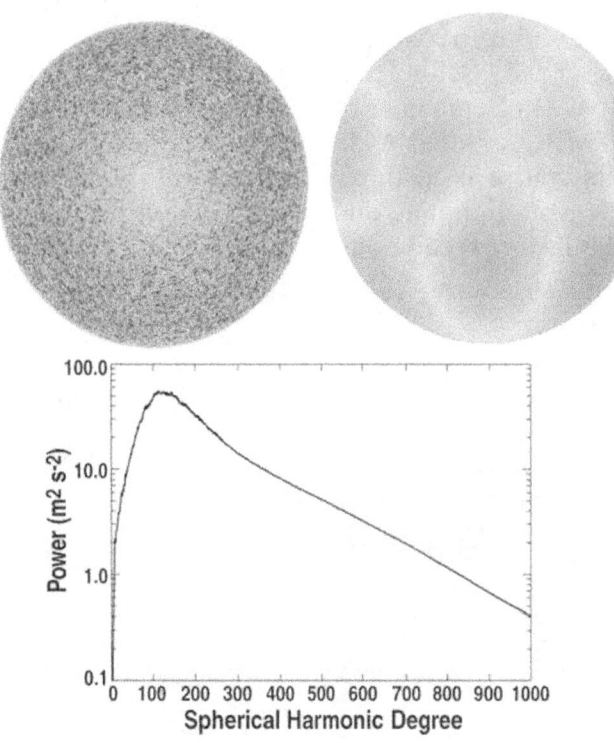

Doppler velocity image (upper left) illustrating flows toward (blue) and away from (red) the observer at the surface of the Sun. A spectrum of convection cell sizes can be derived (bottom), showing a peak corresponding to supergranules. Further spectral analysis provides evidence of instrumental artifacts (upper right), which are of importance in studying weaker convection signals such as those pertaining to giant cells.

Hong Xie

Dr. Xie, of CUA, has been working on data analysis of coronal mass ejections (CMEs), CME-driven shocks, and associated geomagnetic storms. She is currently maintaining the COR1 preliminary CME catalog working with Dr. Chris St. Cyr. Her research interests include the origin, three-dimensional structure, and evolution of CMEs. She has developed a new analytical CME cone model fitting procedure, and incorporated a geometrical flux-rope model for STEREO/COR coronagraph images. Working with Dr. Gopalswamy and Dr. St. Cyr, she studied the effects of solar wind dynamic pressure on large geomagnetic storms and the resulting paper was published in *Geophysical Research Letters* during 2008. She has also analyzed and compared different properties of radio-loud CME-driven shocks and radio-quiet CME-driven shocks. By using flux-rope models and numerical simulations with ENLIL+Cone model, she has investigated 3-D structure and kinematic evolution of the May-13-2005 CME and corresponding magnetic cloud, and the results have been presented at the SHINE workshop and IAU 257 symposium.

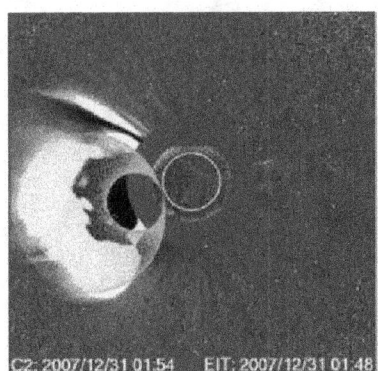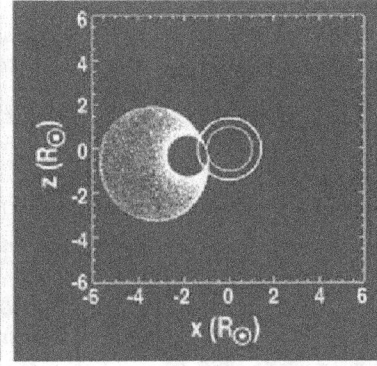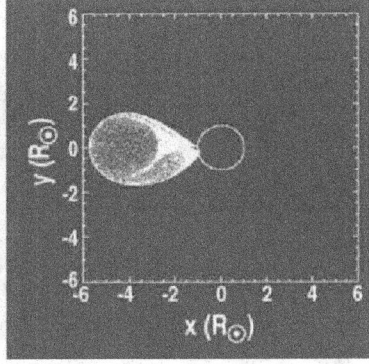

Broadside morphology (middle) and edge on morphology (right) of the flux rope model (courtesy of Dr. Krall's flux-rope model code) for the 2007 December 31 CME. (Left) Flux-rope model superposed with LASCO C2 and EIT 195 at 01:54 UTC, where green curves outline the profile of the flux rope. Note that the streamer deflection (due to the CME-driven shock) above the CME is not included as part of the CME.

C. Alex Young

Dr. Young, of ADNET Systems Inc., works to develop signal and image processing methods and software to facilitate a more complete extraction of scientific information from solar physics data. This is to aid both the community as a whole, as well as his own research into the prediction and understanding of dynamic phenomena in the solar corona such as solar flares and coronal mass ejections. Dr. Young has recently been awarded a NASA grant to develop tools and methods for the analysis of STEREO image data. He was also awarded a NASA grant to support the 4th workshop of the Solar Image Processing Workshop series, held October 2008 in Baltimore, Maryland. This workshop brought together over 70 scientists and image processing experts from around the world. In addition, Dr. Young was a guest editor of a *Solar Physics Journal* special issue in April 2008, which contained papers from some of the material presented at the 3rd Solar Image Processing Workshop. Over 2008, he published several journal papers on his work. His work has included educational outreach both in the form of mentoring several undergraduate and graduate students, and giving interviews for NASA TV and the *Discovery Channel*.

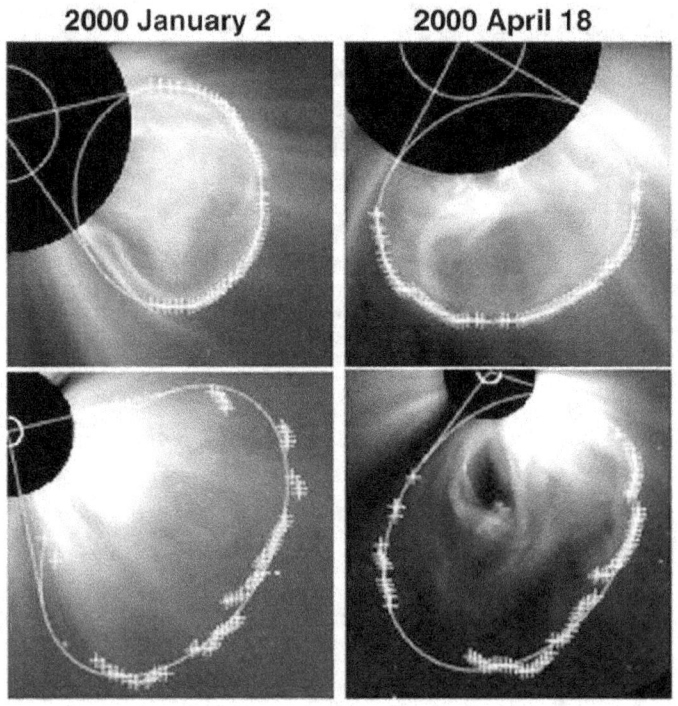

Sample fits to the leading edge of Coronal Mass Ejections (CMEs) observed by the Large Angle and Spectrometric Coronograph (LASCO) instrument aboard SOHO. These fits were used to study the dynamics of the CMEs. This figure was part of the published results from the Ph.D. work of one of Dr. Young's students using some of the tools Dr. Young developed.

Seiji Zenitani

Dr. Zenitani, an astrophysicist, joined Code 674 in November 2006. Currently, he is an NPP researcher at the Oak Ridge Associated Universities (ORAU). He has been doing research on the mechanism of magnetic reconnection in space and astrophysical plasmas, by means of supercomputer simulations. He investigated the basic behaviors of current sheets in astrophysical relativistic electron-positron plasmas.

Working with Drs. Hesse and Klimas, he studied the dissipation mechanism of magnetic reconnection in ion-electron plasmas and in relativistic pair plasmas, and he discovered the Weibel instability in the context of magnetic reconnection (see the figure below).

He also gave several invited talks on the relevant work.

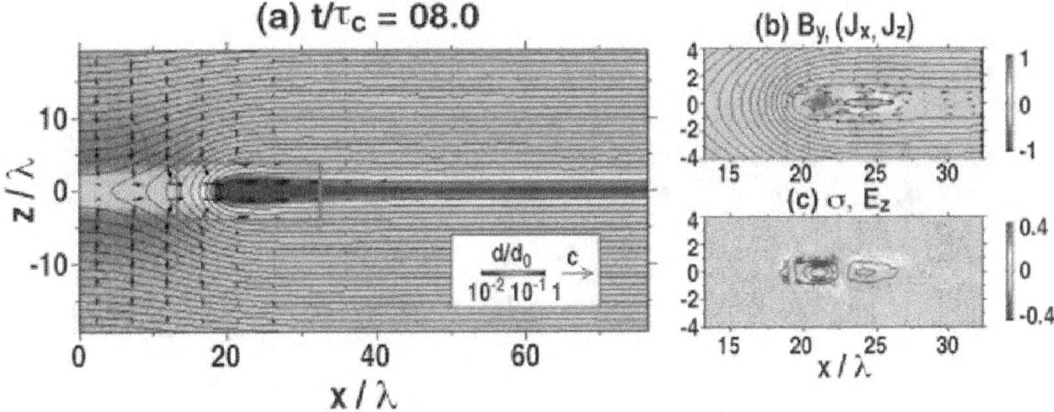

The Weibel instability in the reconnection outflow region. Left: large-scale structure of magnetic reconnection. Right: field perturbations by the Weibel instability.

APPENDIX 2: HSD PUBLICATIONS AND PRESENTATIONS

In FY08, HSD has published a total of 213 papers in a wide variety of scientific journals and proceedings, including two papers in both Nature and Science. Of these, 44% (93) have been with HSD scientists as first author. HSD has submitted or has in press a further 70 articles (28 of which are first author publications). A comprehensive list of these papers is listed in the next two sections of this appendix.

The HSD group gave a total 162 presentations (talks and posters) at a total of 65 different science meetings spread all across the US and around the World. The bulk of the HSD presentations were given at the AGU Fall, AGU/SPD Spring, EGU, and Space Weather Workshop meetings.

These figures may include some publications from before October 2007 but probably overall represent a considerable underestimate of the body of work published and presented by the HSD group at GSFC.

Journal Articles

Abdo, A.A., B. Allen, et al., Discovery of TeV gamma-ray emission from the Cygnus region of the Galaxy, *Astrophys. J.*, **658**(1), L33–L36, 2007.

Acuna, M.H., D. Curtis, J.L. Scheifele, C.T. Russell, P. Schroeder, A. **Szabo**, and J.G. Luhmann, The STEREO/IMPACT magnetic field experiment, *Space Sci. Rev.*, **136**, 203–226, 2008.

Angelopoulos, V., J.P. McFadden, D. Larson, C. W. Carlson, S.B. Mende, H. Frey, T. Phan, D.G. **Sibeck**, K.-H. Glassmeier, U. Auster, E. Donovan, I.R. Mann, I.J. Rae, C.T. Russell, A. Runov, X.-Z. Zhou, and L. Kepko, Tail reconnection triggering substorm onset, *Science,* **321,** 931–934, 2008.

Angelopoulos, V., D. **Sibeck**, C. W. Carlson, J. P. McFadden, D. Larson, R. P. Lin, J. W. Bonnell, F. S. Mozer, R. Ergun, C. Cully, K.-H. Glassmeier, U. Auster, A. Roux, O. Le Contel, S. Frey, T. Phan, S. Mende, H. Frey, E. Donovan, C. T. Russell, R. Strangeway, J. Liu, I. Mann, J. Rae, J. Raeder, X. Li, W. **Liu**, H. J. Singer, V. A. Sergeev, S. Apatenkov, G. Parks, M. Fillingim, J. **Sigwarth**, First results from the THEMIS mission, *Space Sci. Rev.,* 10.1007/s11214-008-9378-4, 2008.

Antiochos, S.K., C.R. DeVore, J.T. **Karpen**, Z. Mikic, Structure and dynamics of the Sun's open magnetic field, *Astrophys. J.,* **671,** 936, 2007.

Baker, D.N., A. **Klimas**, and D. Vassiliadis, "Nonlinear dynamics in the Earth's magnetosphere," In: *Nonlinear Dynamics in Geosciences*, A.A. Tsonis and J.B. Elsner, (Eds.), Springer Verlag, 53–67, 2007.

Benson, R.F., Plasma physics using space-borne radio sounding. In: *CP974, Radio Sounding and Plasma Physics*, P. Song, et al., (Eds.), American Institute of Physics, Lowell, Massachusetts, 20–33, 2008.

Benson, R.F., and D. **Bilitza**, New satellite mission with old data: Rescuing a unique data set, *Radio Sci.*, submitted, 2008. Preliminary version published in the *Proc. 2008 International Ionospheric Effects Symposium,* Radio Propagation Services, Inc., Alexandria, Virginia under the title: New satellite mission with old data: The status of the ISIS data transformation and preservation project.

Berdichevsky, D.B., D.V. Reames, C.-C. **Wu**, R. Schwenn, R.P. **Lepping**, R.J. MacDowall, C.J. Farrugia, J.-L. Bougeret, C. Ng., and A.J. Lazarus, Exploring the global shock scenario at multiple points between the Sun and Earth: The solar transients launched on January 1 and September 23, 1978, *J. Adv. Space Res.,* doi:10.1016/j.asr.2008.03.026, 2008.

Berthelier, J.J., M. Malingre, R. **Pfaff,** E. Seran, R. Pottelette, J. Jasperse, J.-P. Lebreton, and M. Parrot, Lightning-induced plasma turbulence and ion heating in equatorial ionospheric depletions, *Nature Geosci.,* doi:10.1038, 2008.

Bhatia, A.K., and E. Landi, Atomic data and spectral line intensities for Ar XV, *Atomic Data and Nuclear Data Table,* **94**(2), 223–256, 2008.

Bhatia, A.K., Applications of the hybrid theory to the scattering of electrons from He ion and Li doubly ionized ion and resonances in these systems, *Phys. Rev. A,* **77,** 052707 2008.

Bilitza, D., V. Truhlik, P. Richards, T. Abe, and L. Triskova, Solar cycle variation of mid-latitude electron density and temperature: Satellite measurements and model calculations, *J. Adv. Space Res.,* **39**(5), 779–789, doi:10.1016/j.asr.2006.11.022, 2007.

Bilitza, D., and B.W. Reinisch, International Reference Ionosphere 2007: Improvements and new parameters, *J. Adv. Space Res.,* **42**(4), 599–609, doi:10.1016/j.asr.2007.07.048, 2008.

Bilitza, D., B. Reinisch, and J. Lastovicka, Progress in observation-based ionospheric modeling, *Space Weather,* **6**, S02002, doi:10.1029/2007SW000359, 2008.

Bilitza, D., Evaluation of the IRI-2007 model options for the topside electron density, *J. Adv. Space Res.,* 2008.

Bilitza, D., "The importance of bottomside and topside sounding measurements for the development of IRI." In: Radio Sounding and Plasma Physics, American Institute of Physics, *AIP Conf. Proc.,* **974**, 9–19, doi:10.1063/1.2885039, 2008.

Bilitza, D., and B.W. Reinisch, International Reference Ionosphere 2007: Improvements and new parameters, *J. Adv. Space Res.*, **42**(4), 599–609, doi:10.1016/j.asr.2007.07.048, 2008.

Birn, J., J.E. Borovsky, and M. **Hesse**, Properties of asymmetric magnetic reconnection, *Phys. Plasmas*, **15,** 032101, 2008.

Boardsen, S.A., J.L. Green, and B.W. Reinisch, Comparison of kilometric continuum latitudinal radiation patterns with linear mode conversion theory, *J. Geophys. Res.*, **113**, A012919, doi:10.1029/2007JA012319, 2008.

Boardsen, S.A., B.J. Anderson, H. Korth, and J.A. **Slavin**, Ultra-low frequency wave observations by MESSENGER during its January 2008 flyby through Mercury's magnetosphere, *Eos Trans. AGU,* **89**(23), Jt. Assem. Suppl. Abstract U24A-06, 2008.

Breech, B., W.H. Mattaeus, J. Minnie, J.W. Bieber, S. Oughton, C.W. Smith, and P.A. Isenberg, Turbulence transport throughout the heliosphere, *J. Geophys. Res.*, **113**, A08105, 2008.

Breech, B., L. Pollock, and J. Cavazos, RUGRAT: Runtime Test Case Generation using dynamic compilers, *Proc. Intl. Symp. Software Reliability Engineering,* 2008.

Brosius, J.W., D.M. **Rabin**, R.J. **Thomas**, Rapid cadence EUNIS-06 observations of a He II transient brightening in the quiet Sun, *Astrophys. J.,* **682,** 630–637, 2008.

Brosius, J.W., D.M. **Rabin**, R.J. Thomas, and E. Landi, Analysis of a solar coronal bright point extreme ultraviolet spectrum from the EUNIS sounding rocket instrument, *Astrophys. J.*, **677,** 781–789, 2008.

Brosius, J.W., D.M. **Rabin**, R.J. **Thomas**, Analysis of a bright point spectrum from the Extreme Ultraviolet Normal Incidence Spectrograph (EUNIS) Sounding Rocket Instrument, *Eos Trans. AGU,* **88**(52), Fall Meet. Suppl., Abstract SH21B-04, 2007.

Brosius, J.W., D.M. **Rabin**, R.J. **Thomas**, EUNIS-06 Rapid Cadence Observations of a He II Transient Brightening in the Quiet Sun, *Eos Trans. AGU,* **89**, Spring Meet. Suppl., Abstract SP51A-06, 2008.

Burlaga, L.F., N.F. Ness, M.H. **Acuna**, R.P. **Lepping**, J.E.P. Connerney, and J.D. Richardson, Magnetic fields at the solar wind termination shock, *Nature,* **454,** doi:10.10.1038/nature07029, 2008.

Burlaga, L.F., and A.F.-**Viñas**, Tsallis distribution functions in the solar wind: Magnetic field and velocity observations, in complexity, metastability and nonextensivity, *AIP Conf. Proc.,* S. Abe, H. Hermann, P. Quarati, A Rapisarda, and C. Tsallis (Eds.), 259–266, 2007.

Buzulukova, N., M.-C. **Fok**, T. **Moore**, and D. Ober, Generation of plasmaspheric undulations, *Geophys. Res. Lett.*, **35,** 2008.

Buzulukova, N., V. **Vovchenko,** Modeling of proton nose structures in the inner magnetosphere with a self-consistent electric field model, *J. Atmos. Sol. Terr. Phys.,* **70,** 503–510, 2008.

Cameron, R., L. Gizon, T.L. **Duvall**, Jr., Helioseismology of sunspots: Confronting observations with three-dimensional MHD simulations of wave propagation, *Sol. Phys.,* **251,** 291–308, 2008.

Cassidy, T.A., R.E. Johnson, M.A. McGrath, M.C. Wong, and J.F. **Cooper**, The spatial morphology of Europa's near-surface O_2 atmosphere, *Icarus*, **191,** 755–764, 2007.

Chaston, C.C., M. Wilbur, F.S. Mozer, M. Fujimoto, M.L. **Goldstein**, M. **Acuña**, H. Rème, A. Fazakerley, Mode conversion and anomalous transport in Kelvin-Helmholtz vortices and kinetic Alfvén waves at the Earth's magnetopause, *Phys. Rev. Lett.,* **99,** 175004. 2007

Chen, S.-H., and T.E. **Moore**, Ionospheric ions in the near-Earth magnetotail, *J. Geophys. Res.*, **113,** A08232, doi:10.1029/2007JA012816, 2008.

Coates, A.J., F.J. Crary, G.R. Lewis, D.T. Young, J.H. Waite, Jr.; E.C. **Sittler**, Jr. Discovery of heavy negative ions in Titan's ionosphere, *Geophys. Res. Lett.,* **34,** L22103, doi:10.1029/2007GL030978, 2007.

Coates, A.J., F.J. Crary, D.T. Young, K. Szego, C.S. Arridge, Z. Bebesi, E.C. **Sittler,** Jr., R.E. **Hartle**, and T.W. Hill, Ionospheric electrons in Titan's tail: Plasma structure during Cassini T9 encounter, *Geophys. Res. Lett.,* **34,** L24S05, 2007.

Collado-Vega, Y.M., R.L. Kessel, X. Shao, and R.A. Boller, MHD flow visualization of magnetopause boundary region vortices observed during high-speed streams, J. *Geophys. Res.,* **112,** A06213, doi:10.1029/2006JA012104, 2007.

Collier, M.R., R.P. **Lepping**, and D. Berdichevsky, A statistical study of interplanetary shocks and pressure pulses internal to magnetic clouds, *J. Geophys. Res.*, **112,** A06102, doi:101029/2006JA011714, 2007.

Collier, M.R., D.A. **Roberts**, and A.F. **Viñas**, Acoustic kappa-density fluctuation waves in suprathermal kappa function fluids, *Adv. Space Res.,* **1704,** doi:10.1016/j.asr.2007.10.017, 2008.

Collier, M.R., H.K. Hills, and T.J. **Stubbs**, Lunar surface potential changes possibly associated with traversals of the Bow shock, *NLSI Lunar Sci. Conf.*, #2104, 2008.

Collier, M.R., J. Carter, T. Cravens, H.K. Hills, K. Kuntz, F.S. Porter, A. Read, I. Robertson, S. Sembay, S.L. Snowden, T. J. **Stubbs**, and P. Travnicek, The Lunar X-ray Observatory (LXO)/Magnetosheath Explorer in X-rays (MAGEX), *NLSI Lunar Sci. Conf.*, #2082, 2008.

Conlon, P.A., P.T. Gallagher, R.T.J. McAteer, J. **Ireland**, C.A. Young, P. Kestener, and K. Maguire, Multifractal properties of evolving active regions, *Sol. Phys.*, **248**, 297, 2008.

Cooper, P. D., and J. F. Cooper, Water, Water Everywhere... and Oxidants too! *Eos Trans. AGU*, **88**(52), Fall Meet. Suppl., Abstract P32B-01, 2007 (invited).

Cooper, J.F., T.P. Armstrong, M.E. Hill, N. **Lal**, R.E. **McGuire**, R.B. McKibben, T.W. Narock, A. **Szabo**, and C. Tranquille, Virtual Energetic Particle Observatory for the Heliospheric Data Environment, *Eos Trans. AGU*, **88**(52), Fall Meet. Suppl., Abstract SH51A-0251, 2007.

Correira, J., A.C. Aikin, J.M. Grebowsky, W.D. **Pesnell**, and J.P. Burrows, Seasonal variations of magnesium atoms in the mesosphere-thermosphere, *Geophys. Res. Lett.*, **35**, L06103, doi:10.1029/2007GL033047, 2008.

Coustenis, A., S.K. Atreya, T. Balint, R.H. Brown, M.K. Dougherty, F. Ferri, M. Fulchignoni, D. Gautier, R.A. Gowen, C.A. Griffith, L.I. Gurvits, R. Jaumann, Y. Langevin, M.R. Leese, J.I. Lunine, C.P. McKay, X. Moussas, I. Müller-Wodarg, F. Neubauer, T.C. Owen, F. Raulin, **E.C. Sittler**, F. Sohl, C. Sotin, G. Tobie, T. Tokano, E.P. Turtle, J.-E. Wahlund, J.H. Waite, K.H. Baines, J. Blamont, A.J. Coates, I. Dandouras, T. Krimigis, E. Lellouch, R.D. Lorenz, A. Morse, C.C. Porco, M. Hirtzig, J. Saur, T. Spilker, J.C. Zarnecki, E. Choi, N. Achilleos, R. Amils, P. Annan, D.H. Atkinson, Y. Bénilan, C. Bertucci, B. Bézard, G.L. Bjoraker, M. Blanc, L. Boireau, J. Bouman, M. Cabane, M.T. Capria, E. Chassefière, P. Coll, M. Combes, J.F. **Cooper**, et al., TandEM: Titan and Enceladus Mission, *Experimental Astron.*, Online, Springer Netherlands, 54 pp., doi:10.1007/s10686-008-9103-z, 2008.

Croskey, C.L., J.D. Mitchell, R.A. **Goldberg**, and F.J. Schmidlin, An instrumented Loki-Dart payload for measuring mesospheric plasma density variations, *Proc.18th ESA Symp. European Rocket and Balloon Programmes and Related Research,* June 3–7, 2007, Visby, Sweden, ESA-SP647, 241–246, 2007.

Delory, G.T., W.M. Farrell, J.S. Halekas, T.J. **Stubbs**, R.P. Lin, S. Bale, and R. Vondrak, The dynamic lunar environment: Plasmas, neutrals, and dust, *NLSI Lunar Sci. Conf.*, #2086, 2008.

DeMajistre, R., L.J. Paxton, and D. **Bilitza**, Comparison of ionospheric measurements made by digisondes with those inferred from ultraviolet airglow, *Adv. Space Res.*, **39**(5), 918–925, doi:10.1016/j.asr.2006.09.037, 2007.

DeVore, C.R., and S.K. **Antiochos**, Homologous confined filament eruptions via magnetic breakout, *Astrophys. J.,* **680,** 740, 2008.

Eastwood, J.P., D.G. **Sibeck**, V. Angelopoulos, T.-D. Phan, S.D. Bale, J.P. McFadden, C.M. Cully, S.B. Mende, D. Larson, S. Frey, C.W. Carlson, K.-H. Glassmeier, H.U. Auster, A. Roux, and O. Le Contel, THEMIS observations of a hot flow anomaly: Solar wind, magnetosheath and ground-based measurements, *Geophys. Res. Lett.,* **35,** 10.1029/2008GL033475, 2008.

Ebihara, Y., M.-C. **Fok**, J.B. Blake, and J.F. Fennell, Magnetic coupling of the ring current and the radiation belt, *J. Geophys. Res.,* **113,** A07221, doi:10.1029/2008JA013267, 2008.

Eichstedt, J., W.T. **Thompson**, O.C. **St. Cyr**, STEREO ground segment, science operations, and data archive, *Space Sci. Rev.,* **136,** 605–626, 2008.

Eriksson, S., J.T. Gosling, D. Krauss-Varban, T.D. Phan, L.M. Blush, K. Simunac, J.G. Luhmann, C.T. Russell, A.B. Galvin, A. **Szabo**, and M.H. **Acuna**, Shear-flow effects on magnetic field configuration within a solar wind reconnection exhaust, *Geophys. Res. Lett.*, doi:10.1029/2008GL033332, 2008.

Fairfield, D.H., S. Wing, P.T. Newell, J.M. Ruohoniemi, J.T. Gosling, and R.M. Skoug, Polar rain gradients and field-aligned polar cap potentials, *J. Geophys. Res.,* **113,** A10203, doi:10.1029/2008JA013437, 2008.

Farrell, W.M., T.J. **Stubbs**, G.T. Delory, R.R. **Vondrak**, M.R. Collier, J.S. Halekas, and R.P. Lin, Concerning the dissipation of electrically charged objects in the shadowed lunar polar regions, *Geophys. Res. Lett.,* **35,** L19104, doi:10.1029/2008GL034785, 2008.

Farrell, W.M., T.J. **Stubbs**, J.S. Halekas, G.T. Delory, M.R. Collier, R.R. **Vondrak**, and R.P. Lin, Loss of solar wind plasma neutrality and affect on surface potentials near the lunar terminator and shadowed polar regions, *Geophys. Res. Lett.,* **35,** L05105, doi:10.1029/2007GL032653, 2008.

Fludra, A., and J. **Ireland**, Radiative and magnetic properties of solar active regions I. Global magnetic field and EUV line intensities, *Astron. Astrophys,.* **483,** 609, 2008.

Fok, M.-C., R.B. Horne, N.P. Meredith, and S.A. Glauert, The radiation belt environment model: Application to space weather nowcasting, *J. Geophys. Res.,* **113,** A03S08, doi:10.1029/2007JA012558, 2008.

Forbes, K., and O.C. **St. Cyr**, Solar Activity and Economic Fundamentals: Evidence from 12 Geographically Disparate Power Grids, Space Weather, doi:10.1029/2007SW000350, 2008.

Fung, S.F., and X. Shao, Specification of multiple geomagnetic responses to variable solar wind and IMF input, *Ann. Geophys.*, **26,** 639–652, 2008.

Fung, S.F., Radio plasma imager and measurement of magnetospheric field-aligned electron density, in radio sounding and plasma physics, *AIP Proc. 974,* P. Song, J. Foster, M. Mendillo, and D. **Bilitza**, (Eds.), 97–110, 2008.

Gilbert, H.R., A.G. Daou, D. Young, D. Tripathi, and D. Alexander, The filament-Moreton wave interaction of 2006 December 06, *Astrophys. J.,* **685,** 629, 2008.

Goldberg, R.A., A.G. **Feofilov**, A.A. **Kutepov**, W.D. **Pesnell**, and F.J. Schmidlin, Re-evaluation of SABER temperatures triggered by comparison with MaCWAVE MET rocket measurements, *Proc. 18th ESA Symp. European Rocket and Balloon Programmes and Related Research,* June 3–7, 2007, Visby, Sweden, European Space Agency SP-647, 297–300, 2007.

Goossens, M., I. Arregui, J.L. Ballester, and T.J. **Wang**, Analytic approximate seismology of transversely oscillating coronal loops, *Astron. Astrophys,* **484,** 851–857, 2008.

Gopalswamy, N., S. Yashiro, H. **Xie**, S. Akiyama, E. Aguilar-Rodriguez, M.L. **Kaiser**, R.A. Howard, and J.-L. Bougeret, Radio-quiet fast and wide coronal mass ejections, *Astrophys. J.,* **674,** 560–569, 2008.

Gopalswamy, N., S. Akiyama, S Yashiro, G. Michalek, and R.P. **Lepping**, Solar sources of geospace consequences of interplanetary magnetic clouds observed during Solar Cycle 23, *J. Atmos. and Solar Ter. Phys.*, **70**(2-4), 245–253, doi:10.1016/j.jastp.2007.08.070, 2007.

Gosling, J.T., and A. **Szabo**, Bifurcated current sheets produced by magnetic reconnection in the solar wind, *J. Geophys. Res.,* **113,** A10103, /2008JA013473, 2008.

Gosling, J.T., T.D. Phan, R.P. Lin, and A. **Szabo**, Prevalence of magnetic reconnection at small field shears in the solar wind, *Geophys. Res. Lett.*, doi:10.1029/2007GL030706, **34,** L15110, 2007.

Gudipati, M.S., L.J. Allamandola, J.F. **Cooper**, S.J. Sturner, and R.E. Johnson, Consequence of Electron Mobility in Icy Grains on Solar System Objects, *Eos Trans. AGU,* **88**(52), Fall Meet. Suppl., Abstract P53B-1248, 2007.

Halekas, J.S., G.T. Delory, R.P. Lin, T.J. **Stubbs**, and W.M. Farrell, Lunar prospector observations of the electrostatic potential of the lunar surface and its response to incident currents, *J. Geophys. Res.*, **113,** A09102, doi:10.1029/2008JA013194, 2008.

Halekas, J.S., G.T. Delory, T.J. **Stubbs**, W.M. Farrell, and R.P. Lin, The dynamic plasma and electric field environment near the lunar terminator and polar regions, *NLSI Lunar Sci. Conf.*, #2036, 2008.

Halekas, J.S., G.T. Delory, T.J. **Stubbs**, W.M. Farrell, R.P. Lin, Lunar surface charging: magnitude and implications as a function of space and time, *Lunar Planet. Sci. Conf. XXXIX*, #1365, 2008.

Harvey, C.C., M. Gangloff, T. King, C.H. Perry, D.A. **Roberts**, and J. Thieman, Recent developments towards a Solar System Virtual Observatory, *Earth Science Informatics*, doi:10.1007/s12145-008-0008-1, 2008.

Hesse, M., and S. **Zenitani**, Dissipation in relativistic pair-plasma reconnection, *Phys. Plasmas*, **14**, 112102, 2007.

Hewett, R.J., P.T. Gallagher, R.T.J. McAteer, C.A. Young, J. **Ireland**, P.A. Conlon, and K. Maguire, Multiscale analysis of active region evolution, *Sol. Phys.*, **248**, 311, 2008.

Hirzberger, J., L. Gizon, S.K. Solanki, and T.L. **Duvall**, Jr., Structure and evolution of supergranulation from local helioseismology, *Sol. Phys.*, **251**, 417–437, 2008.

Howard, R.A., J.D. Moses, A. Vourlidas, J.S. Newmark, D.G. Socker, S.P. Plunkett, C.M. Korendyke, J.W. Cook, A. Hurley, J.M. **Davila**, W.T. **Thompson**, O.C. **St Cyr**, E. Mentzell, K. Mehalick, J.R. Lemen, J.P. Wuelser, D.W. Duncan, T.D. Tarbell, C.J. Wolfson, A. Moore, R.A. Harrison, N.R. Waltham, J. Lang, C.J. Davis, C.J. Eyles, H. Mapson-Menard, G.M. Simnett, J.P. Halain, J.M. Defise, E. Mazy, P. Rochus, R. Mercier, M.F. Ravet, F. Delmotte, F. Auchere, J.P. Delaboudiniere, V. Bothmer, W. Deutsch, D. Wang, N. Rich, S. Cooper, V. Stephens, G. Maahs, R. Baugh, D. McMullin, and T. Carter, Sun–Earth Connection Coronal and Heliospheric Investigation (SECCHI), *Space Sci. Rev.*, **136**, 67–115, 2008.

Hudson, R.L., M.E. Palumbo, G. Strazzulla, M.H. Moore, J.F. **Cooper**, and S.J. Sturner, *Laboratory Studies of the Chemistry of TNO Surface Materials, in The Solar System Beyond Neptune,* A. Barucci, H. Boehnhardt, D.P. Cruikshank, and A. Morbidelli, (Eds.) 507–523, Univ. of Arizona Press, Tucson, 2008.

Hwang, K.-J., R.E. Ergun, L. Andersson, D.L. Newman, and C.W. Carlson, Test particle simulations of the effect of moving DLs on ion outflow in the auroral downward-current region, *J. Geophys. Res.*, **114**, A01308, doi:10.1029/2007JA012640. 2008.

Hysell, D.L., G. Michhue, M.F. Larsen, R. **Pfaff**, M. Nicolls, C. Heinselman, and H. Bahcivan, Imaging radar observations of Farley Buneman waves during the JOULE II experiment, *Ann. Geophys.*, **26**, 1837, 2008.

Ireland, J., C.A. Young, R.T.J. McAteer, C. Whelan, R.J. Hewett, and P.T. Gallagher, Multi-resolution analysis of active region magnetic structure and its correlation with Mt. Wilson classification and flaring activity, *Sol. Phys.*, **252**, 121–137, DOI 10.1007/s11207-008-9233-5, 2008.

Jackiewicz, J., L. Gizon, A.C. Birch, and T.L. **Duvall**, Jr., Time-distance helioseismology: Sensitivity of f-mode travel times to flows, *Astrophys. J.*, **671**, 1051–1064, 2007.

Jelínek, K., Z. Němeček, J. Šafránková, and J. Merka, Influence of the tilt angle on the bow shock shape and location, *J. Geophys. Res.*, **113**, A05220, doi:10.1029/2007JA012813, 2008.

Jess, D.B., D.M. **Rabin**, R.J. **Thomas**, J.W. **Brosius**, M. Mathioudakis, and F.P. Keenan, Transition-region velocity oscillations observed by EUNIS-06, *Astrophys. J.*, **682**, 1363–1369, 2008.

Ji, H., H. Wang, C. Liu, and B.R. **Dennis**, A hard X-ray sigmoidal structure during the initial phase of the 2003 October 29 X10 flare. *Astrophys. J.*, **680**, 734–739, 2008.

Kaiser, M.L., T.A. **Kucera**, J.M. **Davila**, O.C. **St. Cyr**, M. Guhathakurta, and E. **Christian**, The STEREO mission: An introduction, *Space Sci. Rev.*, **136**, 5–16, 2008.

Karpen, J.T., and S.K. **Antiochos**, Condensation formation by impulsive heating in prominences, *Astrophys. J.*, **676**, 658, 2008.

Kataoka, R., and A. **Pulkkinen**, Geomagnetically induced currents during intense storms driven by coronal mass ejections and corotating interacting regions, *J. Geophy. Res.*, **113**, A03S12, doi:10.1029/2007JA012487, 2008.

Keenan, F.P., D.B. Jess, K.M. Aggarwal, R.J. **Thomas**, J.W. **Brosius**, and J.M. **Davila**, Emission lines of Fe X in active region spectra obtained with the Solar Extreme-ultraviolet Research Telescope and Spectrograph, *Mon. Notices Roy. Astron. Soc.*, **389**, 939, 2008.

Keith, W.R., and T.J. **Stubbs**, Identification of spacecraft conjunctions in the cusps, *Adv. Space Res.*, **41**(10), 1562–1570, doi:10.1016/j.asr.2007.09.030, 2008.

Khazanov, G.V., A.A. Tel'nikhin, and T.K. Kronberg, Nonlinear electron motion in a coherent wave packet, *Phys. Plasmas*, **15**, 073506, 2008.

Khazanov, G.V., A.A. Tel'nikhin, and T.K. Kronberg, Dynamic theory of relativistic electrons stochastic heating by whistler mode waves with application to the Earth magnetosphere, *J. Geophys. Res.*, **113**, A03207, doi:10.1029/2007JA012488, 2008.

Khazanov, G.V., K. Gamayunov, D.L. Gallagher, and J.U. Kozyra, Reply to Comment on "A self-consistent model of the interacting ring current ions and electromagnetic ion cyclotron waves, initial results: Waves and precipitation fluxes" and "Self-consistent model of the magnetospheric ring current and propagating electromagnetic ion cyclotron waves: Waves in multi-ion magnetosphere" by Khazanov et al., *J. Geophys. Res.*, **112**, A12215, doi:10.1029/2007JA012463, 2007.

Khazanov, G.V., and K.V. Gamayunov, Effect of electromagnetic ion cyclotron wave normal angle distribution on relativistic electron scattering in outer radiation belt, *J. Geophys. Res.*, **112**, A10209, doi:10.1029/2007JA012282, 2007.

Khazanov, G.V., and K.V. Gamayunov, Effect of oblique electromagnetic ion cyclotron waves on relativistic electron scattering: Combined Release and Radiation Effects Satellite (CRRES)-based calculation, *J. Geophys. Res.*, **112**, A07220, doi:10.1029/2007JA012300, 2007.

King, T., J. **Merka**, R. Walker, S. Joy, and T. Narock, The architecture of a multi-tiered Virtual Observatory—The VMO. *Earth Sci. Informatics,* doi:10.1007/s12145-008-0006-3, 2008.

King, T., T. Narock, R. Walker, J. **Merka**, and S. Joy, A brave new (virtual) world: Distributed searches, relevance scoring and facets. *Earth Sci. Informatics,* doi:10.1007/s12145-008-0002-7

King, T., R. Walker, J. **Merka**, and T. **Narock**, The Virtual Observatory Experience. *Eos Trans. AGU,* **89**(53), Fall Meet. Suppl., Abstract XXXXX, 2008.

Merka, J., T. Narock, and A. Szabo, 2008 VMO/G status report. *HDMC Meeting,* Baltimore, Maryland, June 10–12, 2008.

Kirkwood, S., A. Belova, D. Murtagh, A. Réchou, R. **Goldberg**, and F. Schmidlin, Polar mesocyclones and their extension to the UTLS-A case study using ESRAD, Odin, and MaCWAVE radiosondes, *Proc. 18th ESA Symp. European Rocket and Balloon Programmes and Related Research,* June 3–7, 2007, Visby, Sweden, European Space Agency, SP-647, 585–588, 2007.

Klimas, A., M. **Hesse**, and S. **Zenitani**, Particle-in-cell simulations of collisionless reconnection with open outflow boundaries, *Phys. Plasmas,* **15**, 082102, 2008.

Klimchuk, J.A., S. Patsourakos, and P.A. Cargill, Highly efficient modeling of dynamic coronal loops, *Astrophys. J.,* **682**, 1351–1362, 2008.

Korotova, G.I., D.G. **Sibeck**, and T. Rosenberg, Seasonal dependence of Interball flux transfer events, *Geophys. Res. Lett.,* **35**, 10.1029/2008GL033254, 2008.

Korotova, G.I., D.G. **Sibeck**, T. Rosenberg, V. Petrov, and V. Styazhkin, Interball observations of multiple flux transfer events, *J. Atmos. Sol. Terr. Phys.,* **70**, 10.1016/j.jastp.2007.08.055, 2008.

Koval, A., and A. **Szabo**, Modified "Rankine-Hugoniot" shock fitting technique: Simultaneous solution for shock normal and speed, *J. Geophys. Res.*, **113**, A10110, doi:10.1029/2008JA013337, 2008.

Kucera, T.A., and E. Landi, An observation of low-level heating in an erupting prominence, *Astrophys. J.*, **673**, 611–620, 2008.

Kutepov, A.A., A.G. **Feofilov**, A.S. Medvedev, A.W.A. Pauldrach, and P. Hartogh, Small-scale temperature fluctuations associated with gravity waves cause additional radiative cooling of mesopause the region. *Geophys. Res. Lett.*, doi:10.1029/2007GL032392, 2007.

Kuznetsova, M.M., M. **Hesse**, L. **Rastatter**, A. Taktakishvili, G. Toth, D.L. De Zeeuw, A. Ridley, and T.I. Gombosi, Multiscale modeling of magnetospheric reconnection *J. Geophys. Res.—Space Phys.*, **112** (A10), A10210, 2007.

Landi, E., and A.K. **Bhatia**, Atomic data and spectral line intensities for S XIII, *Atomic Data and Nuclear Data Tables*, **94**(1), 1–37, 2008.

Le, G., Y. Zheng, C.T. Russell, R.F. **Pfaff**, J.A. **Slavin**, N. Lin, F. Mozer, G. Parks, M. Wilber, S.M. Petrinec, E.A. Lucek, and H. Rème, Flux transfer events simultaneously observed by Polar and Cluster: Flux rope in the subsolar region and flux tube addition to the polar cusp, *J. Geophys. Res.*, **113**, A01205, doi:10.1029/2007JA012377, 2008.

Lee, E., N. Lin, F. Mozer, M. Wilber, E. Lucek, I. Dandouras, H. Rème, J.B. Cao, P. Canu, N. Cornilleau-Wehrlin, P. Décréau, M.L. **Goldstein**, and P. Escoubet, "Density holes in the upstream solar wind." In: *Turbulence and Nonlinear Processes in Astrophysical Plasmas*, D. Shaikh and G.P. Zank, (Eds.), 6th Annual International Astrophysics Conference, APS, Oahu, Hawaii, March 16–22, ISBN: 978-0-7354-0443-4, 2007.

Lepping R.P., T.W. **Narock**, and H. Chen, Comparison of magnetic field observations of an average magnetic cloud with a simple force free model: The importance of field compression and expansion, *Ann. Geophys.*, **25**, 2641–2648, 2007.

Lepping, R.P., and C.-C. **Wu**, On the variation of interplanetary magnetic cloud type through solar cycle 23: WIND events, *J. Geophys. Res.*, **112**, A10103, doi:10.1029/2006JA012140, 2007.

Lepping, R.P., C.-C. **Wu**, N. **Gopalswamy**, and D.B. Berdichevsky, Average thickness of magnetosheath upstream of magnetic clouds at 1 AU vs. solar longitude of source, *Solar Phys.*, doi:10.1007/s11207-007-9111-6, 2008.

Lepping, R.P., C.-C. **Wu**, D. Berdichevsky, and T.J. Ferguson, Estimates of magnetic cloud expansion at 1 AU, *Ann. Geophys.*, **26**, 1919–1933, 2008.

Lepri, S.T., S.K. **Antiochos**, P. Riley, L. Zhao, T. Zurbuchen, Comparison of Heliospheric in-situ Data with the Quasi-Steady Solar Wind Models, *Astrophys. J.*, **674**, 1158, 2008.

Lipatov, A.S., *The Hybrid Multiscale Simulation Technology. An Introduction with Application to Astrophysical and Laboratory Plasmas,* Springer Verlag, New York, 403 pp., 2002.

Liu, J., V. Angelopoulos, D. **Sibeck**, T. Phan, Z.-Y. Pu, J. McFadden, K.-H. Glassmeier, and H.-U. Auster, THEMIS observations of the dayside traveling compression region and flows surrounding flux transfer events, *Geophys. Res. Lett.,* **35,** 10.1029/2008GL033673, 2008.

Liu, W., *Solar Flares as Natural Particle Accelerators: A High-energy View from X-ray Observations and Theoretical Models.* VDM Verlag (Saarbrücken, Germany), ISBN: 978-3-8364-7432-0, 252 pp., 2008.

Liu, W., V. Petrosian, B.R. **Dennis**, and Y.W. Jiang, Double coronal hard and soft X-ray source observed by RHESSI: Evidence for magnetic reconnection and particle acceleration in solar flares, *Astrophys. J.,* **676,** 704–716, 2008.

Liu, W., and V. Petrosian, Double coronal hard X-ray source, *RHESSI Science Nuggets,* Oct. 2007.

López Fuentes, M.C., P. Demoulin, and J.A. **Klimchuk**, Are constant loop widths an artifact of the background and the spatial resolution? *Astrophys. J.,* **673,** 586–597, 2008.

Lui, A.T.Y., D.G. **Sibeck**, T. Phan, V. Angelopoulos, J. McFadden, C. Carlson, D. Larson, J. Bonnell, K.-H. Glassmeier, and S. Frey, Reconstruction of a magnetic flux rope from THEMIS observations, *Geophys. Res. Lett.,* **35,** L17S05, doi:10.1029/2007GL032933, 2008.

Lui, A.T.Y., D.G. **Sibeck**, T. Phan, J.P. McFadden, V. Angelopoulos, and K.-H. Glassmeier, Reconstruction of a flux transfer event based on observations from five THEMIS spacecraft, *J. Geophys. Res.,* **113,** doi:10.1029/2008JA013189, 2008.

Lyatsky, W., and G.V. **Khazanov**, A predictive model for relativistic electrons at geostationary orbit, *Geophys. Res. Lett.,* **35,** L15108, doi:10.1029/2008GL034688, 2008.

Lyatsky, W., and G.V. **Khazanov**, A new polar magnetic index of geomagnetic activity, *Space Weather,* **6,** S06002, doi:10.1029/2007SW000382, 2008.

Lyatsky, W., and G.V. **Khazanov**, Effect of Solar Wind Density on Relativistic Electrons, *Geophys. Res. Lett.,* **35,** L03109, doi:10.1029/2007GL032524, 2008.

Lynch, B.J., S.K. **Antiochos**, C.R. DeVore, J.G. Luhmann, and T.H. Zurbuchen, Topological evolution of a fast magnetic breakout CME in 3-dimensions, *Astrophys. J.,* **683,** 1192, 2008.

Malingre, M., J.-J. Berthelier, R. **Pfaff**, J. Jasperse, and M. Parrot, Lightning-induced lower-hybrid turbulence and trapped Extremely Low Frequency (ELF) electromagnetic waves observed in deep equatorial plasma density depletions during intense magnetic storms, *J. Geophys. Res.*, **113**, A11320, doi:10.1029/2008JA013463, 2008.

Marsh, M.S., J. **Ireland**, and T. **Kucera**, Bayesian analysis of solar oscillations, *Astrophys. J.*, **681**, 672–679, 2008.

Marubashi, K., and R.P. **Lepping**, Long-duration magnetic clouds: A comparison of analyses using torus- and cylindrical-shaped flux rope models, *Ann. Geophys.*, **25**, 2453–2477, 2007.

Matthaeus, W.H., A. Pouquet, P.D. Mininni, P. Dmitruk, and B. **Breech**, Rapid alignment of velocity and magnetic field in magnetohydrodynamic turbulence, *Physical Rev. Lett.*, **100**, 085003, 2008.

McDonald, F.B., E.C. Stone, A.C. Cummings, W.R. Webber, B.C. Heikkila, and N. **Lal**, Anomalous cosmic rays in the distant heliosphere and the reversal of the Sun's magnetic polarity in Cycle 23, *Geophys. Res. Lett.*, **34**(5), CiteID L05105, 2007.

McDonald, F.B., E.C. Stone, A.C. Cummings, W.R. Webber, B.C. Heikkila, N. **Lal**, Anomalous cosmic rays in the distant heliosphere and the reversal of the Sun's magnetic polarity in Cycle 23, *Geophys. Res. Lett.*, **34**, L05105, doi:10.1029/2006GL028932, 2007.

Merka, J., T.W. Narock, and A. Szabo, Navigating through SPASE to heliospheric and magnetospheric data. Earth Science Informatics, **1**, 35–42, doi:10.1007/s12145-008-0004-5, 2008.

Merka, J., D. Merkova, and D. Odstrcil, A step toward data assimilation in solar wind research. *J. Atmos. Sol. Terr. Phys.*, **69**(1), 2,170–2,178, doi:10.1016/j.jastp.2006.07.012, 2007.

Mertens, C.J., J.M. Russell III, M.G. Mlymczak, C.-Y. She, F.J. Schmidlin, R.A. **Goldberg**, M. Lopez-Puertas, P.P. Wintersteiner, R.H. Picard, J.R. Winick, and X. Xu, Kinetic temperature and carbon dioxide from broadband infrared limb emission measurements taken from the TIMED/SABER instrument, *J. Adv. Space Res.*, doi.10.1016/j.asr.2008.04.017, 2008.

Mertens, C., J. Winick, J. Russell III, M. Mlynczak, D. Evans, D. **Bilitza**, and X. Xu, Empirical storm-time correction to the International Reference Ionosphere model E-region electron and ion density parameterizations using observations from TIMED/SABER, *Proc. SPIE, Remote Sensing of Clouds and Atmosphere XII*, **6745**, 67451L, doi:10.1117/12.737318, 2007.

Mierla, M. J. **Davila**, W. **Thompson**, B. Inhester, N. Srivastava, M. **Kramar**, O.C. **St. Cyr**, G. **Stenborg**, R.A. Howard, A Quick method for estimating the propagation direction of coronal mass ejections using STEREO-COR1 images, *Solar Phys.*, **252**(2), doi:10.1007/s11207-008-9267-8, 385–396, 2008.

Moore, T.E., M.-C. **Fok**, D.C. Delcourt, S.P. Slinker, and J.A. Fedder, Global aspects of solar wind—ionosphere interactions, *J. Atmos. Sol. Terr. Phys.*, **69**, 265, doi:10.1016/j.jastp.2006.08.009, 2007.

Moore, T.E., M.-C. **Fok**, D.C. Delcourt, S.P. Slinker, and J.A. Fedder, Plasma plume circulation and impact in an MHD substorm, *J. Geophys. Res.*, **113**, A06219, doi:10.1029/2008JA013050, 2008.

Mueller, D.A.N., and S.K. **Antiochos**, Topologically driven coronal dynamics—a mechanism for coronal hole jets, *Ann. Geophys.*, **26**, 1, 2008.

Narita, Y., K.-H. Glassmeier, M.L. **Goldstein**, and R.A. Treumann, "Cluster observations of shock-turbulence interactions." In: *Turbulence and Nonlinear Processes in Astrophysical Plasmas,* D. Shaikh and G.P. Zank, (Eds.), 6th Annual International Astrophysics Conference, APS, Oahu, Hawaii, March 16–22, 2007, ISBN: 978-0-7354-0443-4, 215–220. 2007.

Narock, T.W., and R.P. **Lepping**, Anisotropy of Magnetic Field Fluctuations in an Average Interplanetary Magnetic Cloud at 1 AU, *J. Geophys. Res.*, **112**, A06108, doi:10.1029/2006JA011987, 2007.

Narock, T.W., A. **Szabo**, and J. **Merka**, Using semantics to extend the space physics data environment. *J. Comp. Geosci.*, doi:10.1016/j.cageo.2007.12.010, 2008.

Narock, T., A. **Szabo**, and J. **Merka**, Opportunities for semantic technologies in the NASA heliophysics data environment. Technical Report SS-08-05. Association for the Advancement of Artificial Intelligence Spring Symposium, Semantic Scientific Knowledge Integration Workshop (AAAI/SSKI08). Stanford University, Stanford, CA. March 26–28, 2008.

Narock, T.W., and T. King, "Developing a SPASE Query Language," *Earth Science Informatics,* doi:10.1007/s12145-008-0007-2, 2008.

Němeček, Z., J. Safránková, J. **Merka**, J. Simunek, and L. Prech, Interball contribution to the high-altitude cusp observations. *Planet. Space Sci.*, **55**(15), 2286–2294, doi:10.1016/j.pss.2007.05.021, 2007.

Ngwira, C.M., A.A. **Pulkkinen**, L.-A. McKinnell, and P.J. Cilliers, Improved modeling of geomagnetically induced currents in the South African power network, *Space Weather*, **6**, S11004, doi:10.1029/2008SW000408, 2008.

Nieves-Chinchilla, T., and A.F. **Viñas**, Solar wind electron distribution functions insidemagnetic clouds, *J. Geophys. Res.*, **113**, A02105, doi:10.1029/2007JA012703, 2008.

Nishino, M.N., M. Fujimoto, T.D. Phan, T. Mukai, Y. Saito, M.M. Kuznetsova, L. **Rastatter**, Anomalous flow deflection at Earth's low-Alfven-mach-number bow shock, *Phys. Rev. Lett.*, **101**(6), 065003, 2008.

Ofman, L., and T.J. **Wang**, Hinode observations of transverse waves with flows in coronal loops, *Astron. Astrophys.*, **482**, L9–L12, 2008.

Ogilvie, K.W., M.A. Coplan, D.A. **Roberts**, and F. Ipavich, Solar wind structure suggested by bimodal correlations of solar wind speed and density between the spacecraft SoHO and Wind, *J. Geophys. Res.*, **112**, A08104, doi:10.1029/2007JA012248, 2007

Øieroset, M., T.D. Phan, D.H. **Fairfield**, J. Raeder, J.T. Gosling, J.F. Drake, and R.P. Lin, The existence and properties of the distant magnetotail during 32 hours of strongly northward interplanetary magnetic field, *J. Geophys. Res.*, **113**, A04206, doi:10.1029/2007JA012679, 2008.

Omidi, N., and D.G. **Sibeck**, Formation of hot flow anomalies and solitary shocks, *J. Geophys. Res.*, **112**, A01203, doi:10.1029/2006JA011663, 2007.

Omidi, N., and D.G. **Sibeck**, Flux transfer events in the cusp, *Geophys. Res. Lett.*, **34**, L04106, doi:10.1029/2006GL028698, 2007.

Owens, M.J., N.U. Crooker, N.A. Schwadron, T.S. Horbury, S. Yashiro, H. **Xie**, O.C. **St Cyr**, and N. **Gopalswamy**, Conservation of open solar magnetic flux and the floor in the heliospheric magnetic field, *Geophys. Res. Lett.*, **35**, L20108, 2008.

Oyama, K.-I., K. Hibino, T. Abe, R. **Pfaff**, and T. Yokoyama, Energetics and structure of the lower E region associated with sporadic-E layer, *Ann. Geophys.*, **26**, 2929, 2008.

Paranicas, C., D.G. Mitchell, S.M. Krimigis, D.C. Hamilton, E. Roussos, N. Krupp, G.H. Jones, R.E. Johnson, J.F. **Cooper**, and T.P. Armstrong, Sources and losses of energetic protons in Saturn's magnetosphere, *Icarus*, **197**, 519–525, 2008.

Pariat, E., S. K. **Antiochos**, R. C. DeVore, A model for solar polar jets, *Astrophys. J.*, **691**, doi:10.1088/0004-637X/691/1/61, 61–74, 2009.

Parks, G.K., E. Lee, A. Teste, M. Wilber, N. Lin, P. Canu, I. Dandouras, H. Rème, S.Y. Fu, M.L. **Goldstein**, Transport of transient solar wind particles in Earth's cusps, *Phys. Plasmas*, Aug. 11, 2008, **15**(8), doi:10.1063/1.2965825, 2008.

Patsourakos, S., E. **Pariat**, A. Vourlidas, S.K. **Antiochos**, J.P. Wuelser, STEREO SECCHI stereoscopic observations constraining the initiation of polar coronal jets, *Astrophys. J., Lett.,* **680,** L73, 2008.

Paulson, D., W.D. **Pesnell**, L. Deming, M. Snow, T. Metcalfe, T. Woods, and B. Hesman, "Chromospheric lines as diagnostics of stellar oscillations." In: Precision Spectroscopy in Astrophysics, *Proc. ESO/Lisbon/Aveiro Conf.,* Aveiro, Portugal, Sept. 11–15, 2006, *Princeton Series in Astrophysics,* N.C. Santos, L. Pasquini, A.C.M. Correia, and M. Romanielleo (Eds.), doi:10.1007/978-3-540-75485-5_79, 311–312, 2008.

Pesnell, W.D., Predictions of solar cycle 24, *Solar Phys.*, doi:10.1007/s11207-008-9252-2, 209–220, 2008.

Peticolas, L.M., N. Craig, T. **Kucera**, et al., The Solar Terrestrial Relations Observatory (STEREO) Education and Outreach (E/PO) Program, *Space Sci. Rev.* **136**, 627–646, 2008.

Pierrard, V., G.V. **Khazanov**, J. Cabrera, and J. Lemaire, Influence of the convection electric field models on predicted plasmapause positionsduring magnetic storms, *J. Geophys. Res.*, **113**, A08212, doi:10.1029/2007JA012612, 2008.

Pfaff, R.F., C. Liebrecht, J.-J. Berthelier, M. Malingre, M. Parrot, and J.-P. Lebreton, "DEMETER satellite observations of plasma irregularities in the topside ionosphere at low, middle, and sub-auroral latitudes and their dependence on magnetic storms." In: *Mid-Latitude Ionospheric Dynamics and Disturbances,* AGU Monograph, P.M. Kintner et al., (Eds.), doi:10.1029/181GM27, 2008.

Pierrard, V., G.V. **Khazanov**, and J. Lemaire, Current-voltage relationship, *J. Atmos. Sol. Terr. Phys.,* **69**(16), 2048–2057, 2007.

Podesta, J.J., D.A. **Roberts**, and M.L. **Goldstein**, Spectral exponents of kinetic and magnetic energy spectra in solar wind turbulence, *Astrophys. J.,* **664**, 543, 2007.

Podesta, J.J., A. Bhattacharjee, B.D.G. Chandran, M.L. **Goldstein**, D.A. **Roberts**, Scale dependent alignment between velocity and magnetic field fluctuations in the solar wind and comparisons to Boldyrev's phenomenological theory, *Proc. IGPP Conf.,* Hawaii, 2008.

Podesta, J.J., M.L. **Goldstein**, and D.A. **Roberts**, Mode decomposition scheme for ideal magnetohydrodynamic plane waves in space-time coordinates, *J. Geophys. Res.*, **112**, doi:10.1029/2006JA012097, 2007.

Porter, F.S., A.F. Abbey, N.P. Bannister, J.A. Carter, M.R. Collier, T.E. Cravens, M. Evans, G.W. Fraser, M. Galeazzi, K. Hills, K. Kuntz, A. Read, I.P. Robertson, S. Sembey, D.G. Sibeck, S. Snowden, T.J. **Stubbs**, P. Travnicek, "The Lunar X-ray Observatory (LXO)." In: Space Telescopes and Instrumentation 2008: Ultraviolet to

Gamma Ray, M.J.L. Turner and K.A. Flanagan, (Eds.), *Proc. SPIE*, **7011**, 70111L, 0277-786X/08/$18, doi:10.1117/12.790182, 2008.

Pulkkinen, A., R. Pirjola, and A. Viljanen, Statistics of extreme geomagnetically induced current events, *Space Weather*, **6**, S07001, doi:10.1029/2008SW000388, 2008.

Reiner, M.J., K.-L. Klein, M. Karlický, K. Jiřička, A. Klassen, M.L. Kaiser, and J.L. Bougeret, Solar origin of the radio attributes of a complex type III burst observed on 11 April 2001, *Solar Phys.*, 249, 337–354, 2008.

Reinisch, B.W., P. Nsumei, X. Huang, and D. **Bilitz**a, Modeling the F2 Topside and Plasmasphere for IRI Using IMAGE/RPI and ISIS Data, *Adv. Space Res.*, **39**(5), 731–738, doi:10.1016/j.asr.2006.05.032, 2007.

Roberts, D.A., J. Giacalone, J.R. Jokipii, M.L. **Goldstein**, and T.D. Zepp, Spectra of polar heliospheric fields and implications for field structure, *J. Geophys. Res.*, **112**, A08103, doi:10.1029/2007JA012247, 2007.

Šafránková, J., Z. Němeček, L. Prech, A. **Koval**, I. Cermak, M. Beranek, G. Zastenker, N. Shevyrev, and L. Chesalin, A new approach to solar wind monitoring, *Adv. Space Res.*, **41**, 153–159, 2008.

Šafránková, J., Z. Němeček, L. Prech, A. Samsonov, A. **Koval**, and K. Andreeova, Interaction of interplanetary shocks with the bow shock, *Planet. Space Sci.*, **55**, 2324–2329, 2007.

Šafránková, J., Z. Němeček, L. Prech, A. Samsonov, A. **Koval**, and K. Andreeova, Modification of interplanetary shocks near the bow shock and through the magnetosheath, *J. Geophys. Res.*, **112**, 8212, 2007.

Sahraoui F., Diagnosis of magnetic structures and intermittency in space plasma turbulence using the technique of surrogate data, *Phys. Rev. E*, **78**, 026402, 2008.

Schindler, K., and M. **Hesse**, Formation of thin bifurcated current sheets by quasi-steady compression, *Phys. Plasmas*, **15**, 042902, 2008.

Schmidt, M., D. **Bilitza**, C.K. Shum, and C. Zeilhofer, Regional 4-D modeling of the ionospheric electron density, *J. Adv. Space Res.*, **42**(4), 782–790, doi:10.1016/j.asr.2007.02.050, 2008.

Sheeley, N.R., Jr., A.D. Herbst, C.A. Palatchi, Y.-M. Wang, R.A. Howard, J.D. Moses, A. Vourlidas, J.S. Newmark, D.G. Socker, S.P. Plunkett, C.M. Korendyke, L.F. Burlaga, J.M. **Davila**, W.T. **Thompson**, O.C. **St Cyr**, R.A. Harrison, C.J. Davis, C.J. Eyles, J.P. Halain, D. Wang, N.B. Rich, K. Battams, E. Esfandiari, and G. Stenborg, Heliospheric images of the solar wind at Earth, *Astrophys. J.*, **675**, 853–862, 2008.

Sheeley, N.R., Jr., A.D. Herbst, C.A. Palatchi, Y.-M. Wang, R.A. Howard, J.D. Moses, A. Vourlidas, J.S. Newmark, D.G. Socker, S.P. Plunkett, C.M. Korendyke, L.F. Burlaga, J.M. **Davila**, W.T. **Thompson**, O.C. **St Cyr**, R.A. Harrison, C.J. Davis, C.J. Eyles, J.P. Halain, D. Wang, N.B. Rich, K. Battams, E. Esfandiari, and G. Stenborg, SECCHI observations of the Sun's garden-hose density spiral, *Astrophys. J.,* **674**, L109–L112, 2008.

Sibeck, D.G., and V. Angelopoulos, THEMIS science objectives and mission phases, *Space Sci. Rev.,* 10.1007/s11214-008-9393-5, 2008.

Sibeck, D.G., M. Kuznetsova, V. Angelopoulos, K.-H. Glassmeier, and J.P. McFadden, Crater FTEs: Simulation results and THEMIS observations, *Geophys. Res. Lett.,* **35**, L17S06, doi:101029/2008GL033568, 2008.

Sibeck, D.G., N. Omidi, I. Dandouras, and E. Lucek, On the edge of the foreshock: model-data comparisons, *Ann. Geophys.,* **26**, 1539–1544, 2008.

Singh, N., G. **Khazanov**, and A. Mukhter, Electrostatic wave generation and transverse ion acceleration by Alfvenic wave components of BBELF turbulence, *J. Geophys. Res.,* **112**(A6), A06210, doi:10.1029/2006JA011933, 2007.

Sittler, E.C., Jr., N. Andre, M. Blanc, M. Burger, R.E. Johnson, A. Coates, A. Rymer, D. Reisenfeld, M.F. Thomsen, A. Persoon, M. Dougherty, H.T. Smith, R.A. Baragiola, R.E. **Hartle**, D. Chornay, M.D. Shappirio, D. Simpson, D.J. McComas, and D.T. Young, Ion and neutral sources and sinks within Saturn's inner magnetosphere: Cassini results, *Planet. Space Sci.,* **56**, 3–18, 2008.

Sittler, E.C., Jr., M. Thomsen, R.E. Johnson, R.E. **Hartle**, M. Burger, D. Chornay, M.D. Shappirio, D. Simpson, H.T. Smith, A.J. Coates, A.M. Rymer, D.J. McComas, D.T. Young, D. Reisenfeld, M. Dougherty, and N. Andre, Erratum to "Cassini observations of Saturn's inner plasmasphere: Saturn orbit insertion results," [*Planet. Space Sci.,* **54** (2006) 1197–1210], *Planet. Space Sci.,* **55**, 2218–2220, 2007.

Slavin, J.A., G. Le, R.J. Strangeway, Y. **Wang,** S.A. Boardsen, M.B. Moldwin, and H.E. Spence, Space Technology 5 multi-point measurements of near-Earth magnetic fields: Initial results, *Geophys. Res. Lett.,* **35**, L02107, doi:10.1029/2007GL031728, 2008.

Slavin, J.A., M.H. Acuna, B.J. Anderson, D.N. Baker, M. Benna, G. Gloeckler, R.E. Gold, G.C. Ho, R.M. Killen, H. Korth, S.M. Krimigis, R.L. McNutt, Jr., L.R. Nittler, J.M. Raines, D. Schriver, S.C. Solomon, R.D. Starr, P. Trávníček, and T.H. Zurbuchen, Mercury's magnetosphere after MESSENGER's first flyby, *Science,* **321**, 85–89, 2008.

Stone, E.C., A.C. Cummings, F.B. McDonald, B.C. Heikkila, N. **Lal**, and W.R. Webber: Voyager 2 finds an asymmetric termination shock and explores the heliosheath beyond, *Nature,* **454**, 71–74, 2008.

Stubbs, T.J., W.M. Farrell, J.S. Halekas, M.W. Collier, G.T. Delory, D.A. Glenar, and R.R. **Vondrak**, A heliophysical monitoring network for the near-surface lunar plasma and radiation environments, *Lunar Planet. Sci. Conf. XXXIX*, **2467,** 2008.

Stubbs, T.J., D.A. Glenar, J.M. Hahn, B.L. Cooper, W.M. Farrell, and R.R. **Vondrak**, Predictions for the lunar horizon glow observed by the Lunar Reconnaissance Orbiter camera, *Lunar Planet. Sci. Conf. XXXIX*, **2378,** 2008.

Stubbs, T.J., D.A. Glenar, R.R. **Vondrak**, J.M. Hahn, B.L. Cooper, and W.M. Farrell, Observing lunar horizon glow with the Hubble Wide Field Camera 3, response to call for white papers on lunar science with HST by the Space Telescope Science Institute, Baltimore, Maryland, 2008.

Sui, L., G.D. **Holman**, and B.R. **Dennis**, Nonthermal X-Ray spectral flattening toward low energies in early impulsive flares. *Astrophys. J.*, **670,** 862–871, 2007.

Sui, L., G.D. **Holman**, and B.R. **Dennis**, A question raised from the observation of dynamic cusp formation: When and where does particle acceleration occur? *Adv. Space Res.*, **41,** 976–983, 2008.

Szego, K., C. Bertucci, A.J. Coates, F. Crary, G. Erdos, R. **Hartle**, E.C. **Sittler**, and D.T. Young, Charged particle environment of Titan during the T9 flyby, *Geophys. Res. Lett.*, **34,** L24S03, 2007.

Taktakishvili A., M.M. Kuznetsova, M. **Hesse**, M.-C. **Fok**, L. **Rastätter**, M. Maddox, A. Chulaki, T.I. Gombosi, D.L. De Zeeuw, Buildup of the ring current during periodic loading-unloading cycles in the magnetotail driven by steady southward interplanetary magnetic field, *J. Geophys. Res.*, **112,** A09203, doi:10.1029/2007JA012317, 2007.

Tan, L.C., D.V. Reames, and C.K. Ng, Bulk flow velocity and first-order anisotropy of solar energetic particles observed on the Wind spacecraft: Overview of three "gradual" particle events, *Astrophys. J.*, **661,** 1297, 2007.

Tan, L.C., D.V. Reames, and C.K. Ng, Ion anisotropy and high-energy variablility of large solar particle events: A comparison study, *Astrophys. J.*, **678,** 1471, 2008.

Thompson, W.T., and **Reginald**, The radiometric and pointing calibration of SECCHI COR1 on STEREO, *Solar Phys.*, **250,** 443–454, 2008.

Vassiliadis, D., A. Pulkkinen, and A. **Klimas**, "Predicting the multifractal geomagnetic field." In: *Nonlinear Dynamics in Geosciences*, A.A. Tsonis and J.B. Elsner (Eds.), Springer Verlag, 581–599, 2007.

Vorotnikov, V.S., C.W. Smith, Q. Hu, A. **Szabo**, R.M. Skoug, and C.M.S. Cohen, Automated shock detection and analysis algorithm for space weather application, *Space Weather*, **6,** S03002, doi:10.1029/2007SW000358, 2008.

Wang, T.J., S.K. Solanki, and M. Selwa, Identification of different types of kink modes in coronal loops: principles and application to TRACE results, *Astron. Astrophys.*, **489**, 1307–1317, 2008.

Webber, W.R., A.C. Cummings, F.B. McDonald, E.C. Stone, B. Heikkila, and N. **Lal**, Galactic cosmic ray H and He nuclei energy spectra measured by Voyagers 1 and 2 near the heliospheric termination shock in positive and negative solar magnetic polarity cycles, *J. Geophys. Res.*, **113**, A10108, 2008.

Webber, W.R., A.C. Cummings, F.B. McDonald, E.C. Stone, B. Heikkila, and N. **Lal**, Passage of a large interplanetary shock from the inner heliosphere to the heliospheric termination shock and beyond: Its effects on cosmic rays at Voyagers 1 and 2, *Geophys. Res. Lett.*, **34**(20), CiteID L20107, 2007.

Webber, W.R., A.C. Cummings, F.B. McDonald, E.C. Stone, B. Heikkila, and N. **Lal**, Temporal and spectral variations of anomalous oxygen nuclei measured by Voyager 1 and Voyager 2 in the outer heliosphere, *J. Geophys. Res.*, **112**, A06105, doi:10.1029/2006JA012207, 2007.

Weimer, D.R., and J.H. **King**, Improved calculations of interplanetary magnetic field phase front angles and propagation delay times, *J. Geophys. Res.*, **113**, A01105, doi:10.1029/2007JA012452, 2008.

Wik, M., A. Viljanen, R. Pirjola, A. **Pulkkinen**, P. Wintoft, and H. Lundstedt, Calculation of geomagnetically induced currents in the 400 kV power grid in southern Sweden, *Space Weather*, **6**, S07005, doi:10.1029/2007SW000343, 2008.

Wu, C.-C., and R.P. **Lepping**, Comparison of the characteristics of magnetic clouds and magnetic cloud-like structures for the events of 1995–2003, *Solar Phys.*, **242**, 159–165, doi:10.1007/s11207-007-0323-6, 2007.

Wu, C.-C., and R.P. **Lepping**, Geomagnetic activity associated with magnetic clouds, magnetic cloud-like structures and interplanetary shocks for the period 1995–2003, *Adv. Space Rev.*, **41**, 335–338, doi:10.1016/j.asr.2007.02.027, 2008.

Xie, H., N. **Gopalswamy**, O.C. **St. Cyr**, and S. Yashiro, Effects of solar wind dynamic pressure and preconditioning on large geomagnetic storms, *Geophys. Res. Lett.*, **35**, L06S08, doi:10.1029/2007GL032298, 2008.

Zenitani, S., and M. Hoshino, Particle acceleration and magnetic dissipation in relativistic current sheet of pair plasmas, *Astrophys. J.*, **670**, 702–726, 2007.

Zenitani, S., and M. Hoshino, The role of the guide field in relativistic pair plasma reconnection, *Astrophys. J.*, **677**, 530–544, 2008.

Zenitani, S., and M. **Hesse**, The role of the Weibel instability at the reconnection jet front in relativistic pair plasma reconnection, *Phys. Plasmas*, **15,** 022101, 2008.

Zenitani, S., and M. **Hesse**, Erratum: The role of the Weibel instability at the reconnection jet front in relativistic pair plasma reconnection, *Phys. Plasmas*, **15,** 089901, 2008.

Zenitani, S., and M. **Hesse**, Self-regulation of the reconnecting current layer in relativistic pair plasma reconnection, *Astrophys. J.,* **684,** 1477–1485, 2008.

Zhang, S.R., J.M. Holt, D. **Bilitza**, T. van Eyken, M. McCready, C. Amory-Macaudier, S. Fukao, and M. Sulzer, Multi-site comparisons between models of incoherent scatter radar and IRI, *J. Adv. Space Res.,* **39**(5), 910–917, doi:10.1016/j.asr.2006.05.027, 2007.

Zheng, Y., A.T.Y. **Lui**, and M.-C. **Fok**, Viewing perspective in energetic neutral atom intensity, *J. Geophys. Res.,* **113,** A09217, doi:10.1029/2008JA013070, 2008.

Zong, Q.-G, H. Zhang, T.A. Fritz, M.L. **Goldstein**, S. Wing, W. Keith, J.D. Winningham, R. Frahm, M.W, Dunlop, A. Korth, P. Daly, H. Rème, A. Balogh, and A.N. Fazakerley, Multiple cusps during an extended northward IMF period with a significant B[y] component, *J. Geophys. Res.*, **113**(A1), A01210, 2008.

Submitted / In Press

Benson, R., and D. **Bilitza**, New satellite mission with old data: rescuing a unique data set, *Radio Sci.*, 2008, submitted.

Bhatia, A.K., and Landi, E., Atomic data and spectral line intensities for Ni XXV, *Atomic Data and Nuclear Data Table*, in press.

Birn, J., and M. **Hesse**, Reconnection in substorms and solar flares: analogies and differences, *Ann. Geophys.*, 2008, submitted.

Boardsen, S.A., B.J. Anderson, M.H. **Acuña**, J.A. **Slavin**, H. Korth, and S.C. Solomon, Narrow-band ultra-low-frequency wave observations by MESSENGER during its January 2008 flyby through Mercury's magnetosphere, *Geophys. Res. Lett.*, doi:10.1029/2008GL036034, 2008, in press.

Brosius, J.W., and G.D. Holman, Observations of the thermal and dynamic evolution of a solar microflare, *Astrophys. J.*, **692,** 2008, in press.

Collier, M.R., and T.J. **Stubbs**, Neutral solar wind generated by lunar exospheric dust at the terminator, *J. Geophys. Res.*, 2008JA013321, 2008, in press.

Cooper, J.F., P.D. Cooper, E.C. **Sittler**, S.J. Sturner, A.M. Rymer, and M.E. Hill, Radiolytic gas-driven cryovolcanism in the outer solar system, *J. Geophys. Res.*, 2007, submitted.

Crooker, N.U., S.W. Kahler, J.T. Gosling, and R.P. **Lepping**, Evidence in magnetic clouds for systematic open flux transport on the Sun, *J. Geophys. Res.*, 2008, accepted.

Dahlburg, R.B., J.-H. Liu, J.A. **Klimchuk**, and G. Nigro, Explosive instability and coronal heating, *Astrophys. J.*, 2008, submitted.

Demoulin, P., and E. **Pariat**, Modelling and observations of photospheric magnetic helicity, *Adv. Space Res.*, 2008, submitted.

DeVore, C.R., S.K. **Antiochos**, Simulating coronal mass ejections in the heliosphere, *Proc. IUA Symp. 257,* 2008, in press.

Feofilov, A.G., A.A. **Kutepov**, M. García-Comas, M. López-Puertas, B.T. Marshall, L.L. Gordley, R.O. Manuilova, V.A. Yankovsky, W.D. **Pesnell**, R.A. **Goldberg**, S.V. Petelina, and J.M. Russell III, SABER/TIMED observations of water vapor in the mesosphere: Retrieval methodology and first results, *J. Atmos. Sol. Terr. Phys.*, 2008, submitted.

Fujii, H.A., T. Watanabe, H. Kojima, K.-I. Oyama, T. **Kusagaya**, Y. Yamagiwa, H. Ohtsu, M. Cho, S. Sasaki, K. Tanaka, J. Williams, B. Rubin, C.L. Johnson, G. Khazanov, J.R. Sanmartin, J.-P. Lebreton, E. van der Heidek, M. Kruijff, F. De Pascale, and P.M. Trivailo, Sounding rocket experiment of bare electrodynamic tether system, *Acta Astronautica,* Elsevier Ltd., 2008, in press.

Gamayunov, K.V., and G.V. **Khazanov**, Crucial role of ring current H+ in electromagnetic ion cyclotron wave dispersion relation: results from global simulations, *J. Geophys. Res.*, 2008, in press.

Gamayunov, K.V., G.V. **Khazanov**, M.W. Liemohn, M-C. **Fok**, and A.A. Ridley, Self consistent model of magnetospheric electric field, Ring 1 current, plasmasphere, and electromagnetic ion Cyclotron 2 waves: Initial results, *J. Geophys. Res.*, 2008, submitted.

Glocer, A., G. Toth, M.-C. **Fok**, T. Gombosi, and M. Liemohn, Integration of the Radiation Belt Environment Model into the Space Weather Modeling Framework, *J. Atmos. Solar-Terrestr. Phys.,* 2008, submitted.

Grebowsky, J.M., R.F. **Benson**, P.A. **Webb**, V. Truhlik, and D. **Bilitza**, Plasmapause-crossing signatures in the ionosphere: A new old data set, *J. Atmos.-Solar Terr. Phys.*, 2008, submitted.

Halekas, J.S., G.T. Delory, R.P. Lin, T.J. **Stubbs**, and W.M. Farrell, Lunar Prospector measurements of secondary electron emission from lunar regolith, *Planet. Space Sci.*, 2008, submitted.

Hesse, M., S. Zenitani, and A. **Klimas**, The structure of the electron outflow jet in collisionless magnetic reconnection, *Phys. Plasmas*, 2008, in press.

Holman, G.D., M.J. Aschwanden, H. Aurass, M. Battaglia, P.C. Grigis, E.P. Kontar, W. **Liu**, P. Saint-Hilaire, and V.V. Zharkova, Implications of X-ray observations for electron acceleration and propagation in solar flares, *Space Sci. Rev.*, 2008, submitted.

Hourcle, J.A., FRBR applied to scientific data, *Proc. ASIS&T*, 2008, submitted.

Huttunen, K.E.J., S.P. Kilpua, A.A. **Pulkkinen**, A.T. Viljanen, and E.I. Tanskanen, Solar wind drivers of large geomagnetically induced currents during the solar cycle 23, *Space Weather*, doi:10.1029/2007SW000374, 2008, in press.

Jelínek, K., Z. Němeček, J. Šafránková, and J. **Merka**, Influence of the tilt angle on the bow shock shape and location. *Geophys. Res. Lett.,* 2008, submitted.

Johnson, R.E., M.H. **Burger**, T.A. Cassidy, F. Leblanc, M. Marconi, W.H. Smyth, Composition and detection of Europa's sputter-induced atmosphere, *Europa*, R. Pappalardo, R. Pappalardo, McKinnon, and K. Khurana (Eds.), University of Arizona Press, 2008, in press.

Karimabadi, H., T.B. Sipes, **Y. Wang**, B. Lavraud, and A. Roberts, Data mining in space physics: 3. Automated detection of flux transfer events using Cluster data, *J. Geophys. Res.*, 2008, submitted.

Khazanov, G.V., A.A. Tel'nikhin, and T.K. Kronberg, "Chaotic Motion of Relativistic Electrons Driven by Whistler Waves," In: *Plasma Physics Book*, 2008, submitted.

Khazanov, G.V., W. Lyatsky, and J.U. Kozyra, Solar cycle effect in relativistic electrons at geostationary orbit, *Geophys. Res. Lett.*, 2008, submitted.

Kirk, M.S., W.D. **Pesnell**, C.A. Young, and S.A. Hess Webber, Automated detection of EUV polar coronal holes during solar cycle 23, *Solar Phys.*, 2008, in press.

Klimchuk, J.A., et al., Commission 10: Solar Activity (2009 Triennial Report), *Trans. IAU*, **XXVIIA**, Reports on Astronomy 2006–2009, K.A. van der Hucht (Eds.) Cambridge University Press, 2008, in press.

Kontar, E.P., J.C. Brown, A.G. Emslie, W. Hajdas, G.D. **Holman**, G.J. Hurford, J. Kasparova, P.C.V. Mallik, A.M. Massone, M.L. McConnell, M. Piana, M. Prato, E.J. Schmahl, and E. Suarez-Garcia, Deducing electron properties from hard X-ray observations, *Space Sci. Rev.*, 2008, submitted.

Landi, E., and A.K. **Bhatia**, Atomic data and spectral line intensities for Ca XVII, *Atomic Data and Nuclear Data Table,* [data calculated for the above ions are added to CHIANTI at NRL], 2008, in press.

Lepping, R.P., T. **Narock**, and C.-C. **Wu**, A scheme for finding the front boundary of an interplanetary magnetic cloud, *Ann. Geophys.*, 2008, submitted.

Liu, R., D. Alexander, H.R. **Gilbert**, Asymmetric eruptive filaments, *Astrophys. J.,* 2008, in press.

Liu, W., V. Petrosian, B.R. **Dennis**, and G.D. **Holman**, Conjugate hard X-ray footpoints in the 2003 October 29 X10 Flare: Unshearing motions, correlations, and csymmetries, *Astrophys. J.,* **693,** March 2009, in press.

Lytskaya, S., Lyatsky, W., and G. V. **Khazanov**, "Relationship between Substorm Activity and Magnetic Disturbances" in Two Polar Caps, *Geophys. Res. Lett.*, 2008, in press, .

Lyatsky, W., and G. V. **Khazanov**, Effect of Geomagnetic Disturbances and Solar Wind Density on Relativistic Electrons at Geostationary Orbit, *Journal of Geophysical Research*, in press, 2008.

Lytskaya, S., W. Lyatsky, and G.V. **Khazanov**, Auroral electrojet al index and magnetic disturbances in two hemispheres, *Geophys. Res. Lett.,* 2008, submitted.

Melrose, D.B., J.A. **Klimchuk**, et al., Division II: Sun and Heliosphere (2009 Triennial Report), *Trans. IAU*, **XXVIIA**, Reports on Astronomy 2006–2009, K.A. van der Hucht (Ed.) Cambridge University Press, 2008, in press.

Narita, Y., K.-H Glassmeier, P.S. Gary, M. **Goldstein**, **F. Sahraoui,** and R. Treumann, Wave number dependence of magnetic turbulence spectra in the solar wind, *Phys. Rev. Lett.,* 2008, submitted.

Němeček, Z., J. Šafránková , A. **Koval** , J. **Merka** , and L. Prech, MHD analysis of propagation of an interplanetary shock across magnetospheric boundaries, *J. Geophys. Res.*, 2008, submitted.

Ogilvie, K.W., D.A. Roberts, A.V. Usmanov, M.Stevens, The relationship between extremely low density events in the solar wind, *J. Geophys. Res.,* 2008, submitted.

Owens, M.J., N.U. Crooker, N.A. Schwadron, T.S. Horbury, S. Yashiro, H. **Xie**, O.C. **St. Cyr**, and N. **Gopalswamy**, Conservation of open solar magnetic flux and the floor in the heliospheric magnetic field, *J. Geophys. Res.*, 2008, in press.

Podesta, J.J., B. Chandran, A. Bhattacharjee D.A. **Roberts**, and M.L. **Goldstein**, Angle between velocity and magnetic field fluctuations in solar wind turbulence, *Astrophys. J.,* 2008, in press.

Paranicas, C., J.F. **Cooper**, H.B. Garrett, R.E. Johnson, and S.J. Sturner. "Europa's Radiation Environment and its Effect on the Surface." In: *Europa*, R. Pappalardo, W.B. McKinnon, and K. Khurana, (Eds.), University of Arizona Press Space Science Series, 2008, in press.

Patsourakos, S., and J.A. **Klimchuk**, Static and impulsive models of solar active regions, *Astrophys. J.,* 2008, in press.

Patsourakos, S. and J.A. **Klimchuk**, Spectroscopic observations of hot lines constraining coronal heating in solar active regions, *Astrophys. J.,* 2008, submitted.

Podesta, J.J., B. Chandran, A. Bhattacharjee D.A. Roberts, and M.L. Goldstein, *Astrophys. J.,* 2008, in press.

Pulkkinen, A., and L. Rastätter, On the minimum variance analysis-based propagation of the solar wind observations, *J. Geophys. Res.*, 2008, submitted.

Pulkkinen, A., A. Taktakishvili, and W. Jacobs, Novel approach to geomagnetically induced current forecasts based on remote solar observations, *Space Weather*, 2008, submitted.

Pulkkinen, A., M. **Hesse**, S. Habib, L. Van der Zel, B. Damsky, F. Policelli, D. Fugate, and W. Jacobs, Solar Shield—forecasting and mitigating space weather effects on high-voltage power transmission systems, *Remote Sensing Applications for Societal Benefits*, S. Habib (Ed.), Springer, New York, 2008, accepted.

Pulkkinen, A., A. Viljanen, and R. Pirjola, Harnessing the celestial batteries, *American J. Phys.*, 2008, submitted.

Raftery, C.L., P.T. Gallagher, R.O. Milligan, and J.A. **Klimchuk**, Multi-Wavelength observations and modeling of a solar flare, *Astron. Astrophys.*, 2008, in press.

Reiner, M.J., K. Goetz, J. Fainberg, M.L. **Kaiser**, M. Maksimovic, B. Cecconi, S. Hoang, S.D. Bale, and J.-L. Bougeret, Multipoint observations of solar Type III radio bursts from STEREO and Wind, *Solar Phys.*, 2008, submitted.

Robertson, I.P., S. Sembay, T.J. **Stubbs**, K. Kuntz, M. Collier, T. Cravens, S. Snowden, H. Hills, F. Porter, P. Travnicek, J. Carter, and A. Read, Solar wind charge exchange observed through the lunar atmosphere, *Geophys. Res. Lett.*, 2008GL035170, 2008, submitted.

Saba, J.L.R., G.L. Slater, and K.T. **Strong**, The rapid onset of solar cycle 23, *Astrophys. J.*, 2008, submitted.

Sahraoui, F., Surrogate data method to analyze structures and intermittency in space plasma turbulence, *Proc. 15th Cluster Workshop,* Tenerife, March 7–15, 2008, 2008, submitted.

Sittler, E.C., et al., "Energy Deposition Processes in Titan's Upper Atmosphere." In: *Titan after Cassini-Huygens,* 2008, submitted.

Strong, K.T., and J.L.R. **Saba**, A New Approach to Solar Forecasting, *Adv. Space Res.*, 2008, accepted.

Taktakishvili, A., M. Kuznetsova, P. **MacNeice**, M. **Hesse**, L. Rastätter, A. **Pulkkinen**, and A. Chulaki, Validation of the coronal mass ejection predictions at the Earth orbit estimated by ENLIL heliosphere cone model, *Space Weather*, 2008, submitted.

Stubbs, T.J., W.M. Farrell, J.S. Halekas, J.K. Burchill, M.R. Collier, R.R. Vondrak, G.T. Delory, and R.F. Pfaff, Dependence of lunar surface charging on ambient solar wind plasma conditions and solar irradiation, *J. Geophys. Res.*, 2008JA013260, 2008, submitted.

Thompson, W.T., 3-D triangulation of a Sun-grazing comet, *Icarus,* 2008, submitted.

Thompson, W.T., Precision effects for solar image coordinates, *Astron. Astrophys.*, 2008, submitted.

Truhlik, V., D. **Bilitza**, and L. Triskova, Latitudinal variation of the topside electron temperature at different levels of solar activity, *J. Adv. Space Res.*, 2008, submitted.

Uritsky, V. M., E. Donovan, A. J. Klimas, and E. Spanswick, Scale-free and scale-dependent modes of energy release dynamics in the nighttime magnetosphere, *Geophys. Res. Lett.*, 2008, in press.

Viljanen, A., A. Pulkkinen, and R. Pirjola, Prediction of the geomagnetic K index based on its previous values, *Geophysica*, 2008, in press.

Viñas, A.F., and C. Gurgiolo, Spherical harmonic analysis of particle velocity distribution function: comparison of moments and anisotropies using cluster data, *J. Geophys. Res.*, 2008, in press.

Wang, Y., G. Le, J.A. **Slavin**, S.A. Boardsen, and R.J. Strangeway, Space Technology 5 measurements of auroral field-aligned current sheet motion, *Geophys. Res. Lett.*, 2008, submitted.

Wik, M., R. Pirjola, H. Lundstedt, A. Viljanen, P. Wintoft, and A. **Pulkkinen**, Comparison of the great space weather events in July 1982 and October 2003 including effects of geomagnetically induced currents on Swedish technical systems, *Ann. Geophys.*, 2008, accepted.

Wu, C.-C., and R.P. **Lepping**, Comparison of the characteristics of magnetic clouds and magnetic cloud-like structures for the events of 1995–2003, *Solar Phys.*, 2008, submitted.

Presentations

Antiochos, S.K., "Solar Reconnection," AGU, San Francisco, Dec. 2007 (invited).

Antiochos, S.K., "A Model for Coronal Hole Jets," AGU, San Francisco, Dec. 2007 (invited).

Antiochos, S.K., "Modeling Solar Flares and CMEs," Space Policy Institute Workshop, Washington, DC, Jan. 2008 (invited).

Antiochos, S.K., "CME Initiation and Heliospheric Implications," 2nd Heliospheric Network Workshop, Kefalonia, Greece, May 2008 (invited).

Antiochos, S.K., "Coronal Heating and Structure," SPD-AGU, Fort Lauderdale, Florida, May 2008 (invited).

Antiochos, S.K., "CME Initiation," 2nd Hinode Science Meeting, Boulder, Colorado, Sept. 2008 (invited).

Benson, R.F., and D. **Bilitza**, "New satellite mission with old data: The status of the ISIS data transformation and preservation project," paper A121 presented at the 12th International Ionospheric Effects Symposium, Alexandria, Virginia, May 13–15, 2008.

Benson, R.F., J.M. Grebowsky, D. **Bilitza**, P.A. **Webb**, and V. Truhlik, "Investigation of the latitude and altitude dependence of large mid-latitude ionospheric electron and ion density gradients associated with outer-plasmaspheric structures," paper GP1-01.13 presented at the XXIX URSI General Assembly, Chicago, Illinois, August 7–16, 2008.

Benson, R.F., M.L. **Adrian**, M.D. Deshpande, W.M. Farrell, S.F. Fung, V.A. **Osherovich**, R.F. **Pfaff,** and D.R. **Rowland**, "Lessons learned from previous space-borne sounders as a guide to future sounder development," paper GH.10, XXIX URSI General Assembly, Chicago, Illinois, held August 7–16, 2008.

Boardsen, S.A., "Search for thermalization of Na+ pickup ions in Mercury's magnetosheath and magnetosphere via observation and hybrid simulation," Heliophysics Science Division Seminar, GSFC, Greenbelt, Maryland, January 11, 2008.

Boardsen, S., B. Anderson, H. Korth, and J. **Slavin**, M. **Acuna**, J. Green, J. Raines, T. Zurbuchen, D. Schriver, and P. Travnicek, "Ultra-low-frequency wave observations by messenger during its January 2008 flyby through Mercury's magnetosphere," 2008 European Planetary Science Congress, Muenster, Germany, September 21–26, 2008.

Brosius, J.W., D.M. **Rabin**, R.J. **Thomas**, D.B. Jess, "The Extreme-Ultraviolet Normal-Incidence Spectrograph (EUNIS) Sounding Rocket Instrument," NAM-2008 Abstract P37/343.

Burger, M.H., "No sodium in the Enceladus plume: Implication for a sub-surface ocean." Polar Gateways Arctic Circle Sunrise 2008, Barrow, Alaska, January 2008.

Burger, M.H., Presentation given at the Physics Department Colloquium, George Washington University, February 2008.

Burger, M.H., Presentation given at Science Night, Parkland Magnet Middle School for Science and Technology, Rockville, Maryland, April 2008.

Burger, M.H., Physics demonstrations given at Fred Ipalook Elementary School, Barrow, Alaska, January 2008.

Chen, P.C., M.E. Van Steenberg, R.J. Oliversen, and D.M. **Rabin**, "Moon dust telescopes, solar concentrators, and structures," presented at the 212th AAS meeting, St. Louis, Missouri, June 1–5, 2008.

Chen, P.C., R.G. Lyon, M.E. Van Steenberg, "Optical design and in situ fabrication of large telescopes on the Moon," paper No. 7010-163, SPIE Conf. on Space Telescopes and Instrumentation, June 23–28, 2008, Marseille, France, NASA Press Release #060408, June 4, 2008.

Cooper, J.F., R.E. **McGuire**, N. **Lal**, D. **Bilitza**, M.E. Hill, T.P. Armstrong, R.B. McKibben, A. **Szabo**, and C. Tranquille, "Virtual Cosmic Ray Observatory (ViCRO)," Paper 0377, *Conf. Proc. 30th International Cosmic Ray Conf.*, Merida, Mexico, July 3–11, 2007.

Cooper, J.F., "Niagara Falls cascade model for interstellar energetic ions in the heliosheath," Paper 0386, *Conf. Proc. 30th International Cosmic Ray Conf.*, Merida, Mexico, July 3–11, 2007.

Cooper, J.F., "Enceladus and the inner magnetosphere of Saturn," Heliophysics Science Division Seminar, NASA Goddard Space Flight Center, Greenbelt, Maryland, November 30, 2007.

Cooper, J.F., "The Global Heliophysics Observatory for IPY-IHY 2007–2009," Polar Gateways Arctic Circle Sunrise 2008 Conference, Barrow, Alaska, January 23–29, 2008.

Cooper, J.F., "Comparative radiation environments of Europa, Enceladus, and Saturn's main rings," poster PS4.1-1TU3P-0568, EGU General Assembly 2008, Vienna, Austria, April 13–18, 2008.

Cooper, J.F., "Polar Gateways to Heliospheric Exploration (POGHEX) for ICESTAR-IHY 2007–2009," talk, ICESTAR/IHY Workshop, EGU General Assembly 2008, Vienna, Austria, April 13–18, 2008.

Cooper, J.F., R.E. **Hartle**, and E.C. **Sittler**, Jr., "Lunar Surface Solar Origins Explorer (LunaSSOX)," contributed mission concept poster, Heliospheric Roadmap Town Hall Meeting, College Park, Maryland, May 19–20, 2008.

Cooper, J.F., "Beyond the heliopause—the very local interstellar radiation environment," talk at Science Working Team Meeting, The Interstellar Boundary Explorer Mission, Boston University, Boston, Massachusetts, June 19–20, 2008.

Cooper, J.F., "Enceladus bidirectional interactions with the Saturn magnetosphere," paper B03-0033-08 (oral), 37th COSPAR Scientific Assembly, Montreal, Canada, July 13–20, 2008.

Cooper, J.F., K. Kauristie, A.T. Weatherwax, G.W. Sheehan, R.W. Smith, I. Sandahl, N. Østgaard, S. Chernouss, B.J. Thompson, L. Peticolas, M.H. Moore, D.A Senske, L.K. Tamppari, and E.M. Lewis, IHY-IPY conference report from Polar Gateways Arctic Circle Sunrise 2008, Paper D12-0018-08 (oral), 37th COSPAR Scientific Assembly, Montreal, Canada, July 13–20, 2008.

Cooper, J.F., E.C. **Sittler**, Jr., R.E. **Hartle**, and S.J. Sturner, "High energy radiation environments of the Jupiter and Saturn systems—impacts and mitigation," Abstract: EPSC2008-A-00118 (accepted, oral), European Planetary Science Congress 2008, Muenster, Germany, September 21–26. 2008.

Dennis, B.R., L.C. Dang, R. Jain, R.A. **Schwartz**, R. Starr, A.K. **Tolbert**, and A. Gopie, "Iron abundance of flare plasma," SPD meeting, Fort Lauderdale, Florida, June 2008.

Dennis, B.R., and R. Pernak, "Determination of HXR Source Sizes with RHESSI Image Reconstruction," RHESSI Science Workshop, AIP, Potsdam, Germany, September 2–6, 2008.

Dennis, B.R., R.D. Starr, A.A. Gopie, R.A. **Schwartz**, and A.K. **Tolbert**, "Abundances of Flare Plasma," RHESSI Science Workshop, AIP, Potsdam, Germany, September 2–6, 2008.

Feofilov, A.G., A.A. **Kutepov**, B.T. Marshall, W.D. **Pesnell**, R.A. **Goldberg**, L.L. Gordley, and J.M. Russell III, "Mesospheric water vapor densities measured by SABER/TIMED," 5th IAGA/ICMA/CAWSES workshop "Long-term changes and trends in the atmosphere," St. Petersburg, Russia, September 9–13, 2008.

Feofilov, A.G., A.A. **Kutepov**, M. García-Comas, M. López-Puertas, B.T. Marshall, L.L. Gordley, R.O. Manuilova, V.A. Yankovsky, W.D. **Pesnell**, R.A. **Goldberg**, S.V.

Petelina, and J.M. Russell III, "SABER/TIMED Mesospheric Water Vapor and Temperature," AGU Joint Assembly, Fort Lauderdale, Florida, 2008.

Feofilov, A.G., and the SABER H$_2$O Team, "Mesospheric H$_2$O Densities Retrieved from SABER/TIMED Measurements," Vol. 10, EGU2008-A-04450, EGU General Assembly, Vienna, Austria, 2008.

Feofilov, B.T., M. Marshall, A.A. Garcia-Comas, M. **Kutepov**, R.O. Lopez-Puertas, V.A. Manuilova, W.D. Yankovsky, R.A. **Pesnell**, R.A. **Goldberg**, L.L. Gordley, S. Petelina, and J.M. Russell III. "Mesospheric water vapor retrieved from SABER/TIMED measurements," paper #SA41A-0291, AGU Fall Meeting, San Francisco, California, 2007.

Feofilov, A.G., B.T. Marshall, M. Garcia-Comas, A.A. Kutepov, M. Lopez-Puertas, R.O. Manuilova, V.A. Yankovsky, W.D. **Pesnell**, R.A. **Goldberg**, L.L. Gordley, S. Petelina, and J.M. Russell III. "Non-LTE model of H$_2$O emissions and its application for SABER/TIMED measurements," 4th International Atmospheric Limb Workshop, Virginia Beach, Virginia, October 29–November 2, 2007.

Fok, M.-C., R.B. Horne, N.P. Meredith, and S.A. Glauert, "Storm-dependent radiation belt dynamics," AOGS 5th Annual Meeting, Busan, Korea, June 2008.

Fok, M.-C., "Understanding the Sun-Earth connection through global imaging," Third International Symposium on KuaFu Project, Kunming, China, September 2008.

Fok, M.-C., and T.E. **Moore**, "Energy and mass transport of magnetospheric plasma during the November 2003 magnetic storm," Huntsville, Alabama, October 2008.

Gamayunov, K.V., G.V. **Khazanov**, M.W. Liemohn, M-C. **Fok**, and A.A. Ridley, "Self-consistent model of magnetospheric electric field, RC and EMIC waves," AGU Fall Meeting, San Francisco, California, December 10–14, 2007.

Goldstein, M.L., A.V. **Usmanov**, and D.A. **Roberts**, "Simulations of solar wind turbulence," ASTRONUM 2008 3rd International Conference on Numerical Modeling of Space Plasma Flows, St. John, US Virgin Islands, June 8–13, 2008.

Gurman, J.B., J.A. **Hourclé**, R.S. Bogart, K. Tian, F. Hill, I. Suárez-Solá, D.M. **Zarro**, A.R. Davey, P.C. Martens, and K. Yoshimura, "Callable Virtual Observatory Functionality: Sample Use Cases," American Geophysical Union, Fall Meeting 2007, abstract #SH51A-0259, 2007.

Gurman, J.B., D.M. **Zarro**, R.S. Bogart, F. Hill, A.R. Davey, P.C. Martens, and the VSO Team, "Still virtual after all these years: Recent developments in the Virtual Solar Observatory," Spring AGU/SPD meeting, abstract SP51B-17, 2008.

Hesse, M., "The structure of the electron outflow jet in collisionless magnetic reconnection," 2008 Cambridge workshop on Magnetic Reconnection, Bad Honnef, Germany, August 2008.

Hesse, M., "The electron diffusion region: Forces and currents," Symposium on the occasion of the 75th birthday of Russell Kulsrud, Princeton University, Princeton, New Jersey, July 2008.

Hesse, M., "On the onset of magnetic reconnection," SHINE/GEM joint workshop, Zermatt, Utah, July 2008.

Hesse, M., "On plasmoid acceleration in the magnetotail," SHINE/GEM joint workshop, Zermatt, Utah, July 2008.

Hesse, M., "The diffusion region and the electron outflow jet," GEM 2008, Zermatt, Utah, June 2008.

Hesse, M., "Real-time and event-based prediction capabilities of modern space science models," Space Weather Workshop, Boulder, Colorado, April 2008.

Hesse, M., "NASA research missions: Applications for space weather forecasting," Air Force Weather Agency planning workshop, Omaha, Nebraska, April 2008.

Hesse, M., "The dissipation mechanism of magnetic reconnection," 2008 US–Japan Reconnection Workshop, Okinawa, Japan, March 2008.

Hesse, M., "Space Weather Models at the CCMC and their capabilities," Fall Meeting of the American Geophysical Union, San Francisco, California, December 2007.

Holland, M., and A. **Pulkkinen**, "Open-source peer-to-peer environment to enable Sensor Web architecture: Application to geomagnetic observations and modeling," paper, Fall AGU 2007, San Francisco, 10–14 December, 2007.

Holman, G., "Observational results from the Ramaty High Energy Solar Spectroscopic Imager (RHESSI) and their implications for flare models," National Research Laboratory, September 2008.

Holman, G.D., L. **Sui**, and Y. Su, "Inferring the Energy Distribution of Accelerated Electrons in Solar Flares from X-ray Observations," Eos, Spring AGU, SH43B-02, 2008.

Holman, G.D., "X-Ray source motions and their implications for flare models," 37th COSPAR Scientific Assembly, 37, 1261, 2008.

Ireland, **J.**, "Correlation of multi-resolution analyses of active region magnetic field structure with flare activity," poster, AGU, San Francisco, California, December 2007.

Ireland, J., "Development of an automated oscillation detection algorithm," poster and talk, SDO-AIA, SDO Science Meeting, Napa, California, March 2008.

Ireland, J., "Helioviewer.org—The inner heliosphere in your browser," poster and talk, SDO-AIA, SDO Science Meeting, Napa, California, March 2008.

Ireland, J., "Multi-scale structure of active regions," NAOJ Seminar, Japan, June 2008.

Ireland, J., "Multi-scale analyses of active region magnetic structure and its correlation with the Mt. Wilson classification and flare activity," University of Central Lancashire Seminar, United Kingdom, July 2008.

Ireland, J., "Automated detection of oscillations in extreme ultraviolet imaging data," poster, ESPM-12, Freiburg, Germany, September 2008.

Ireland, J., "Automated detection of oscillations in extreme ultraviolet imaging data," seminar, Royal Observatory of Belgium, Brussels, Belgium, September 2008.

Jackman, C.H., and W.D. **Pesnell,** "An overview of the impact of energetic particle precipitation (epp) on the mesosphere and stratosphere," International Workshop on Solar Variability, Earth's Climate and the Space Environment, Bozeman, Montana, June 2008.

Jess, D.B., D.M. **Rabin**, R.J. **Thomas**, J.W. **Brosius**, M. Mathioudakis, and F.P. Keenan, "Transition-region velocity oscillations observed by EUNIS-06," NAM-2008 Abstract P31/113, 2008.

Kauristie, K., A. Weatherwax, R. Stamper, J.F. **Cooper**, E. Tanskanen, and the ICESTAR/IHY Team, IPY Project "Heliosphere Impact on Geospace," Science and Outreach, Poster S2.3-P08, Polar Research—Arctic and Antarctic Perspectives in the International Polar Year, SCAR/IASC IPY Open Science Conference, St. Petersburg, Russia, July 8–11, 2008.

Keller, J.W., M.R. Collier, D. Chornay, P. Roz, S. Getty, J.F. **Cooper**, and B. Smith, A New Instrument Design for Imaging Low Energy Neutral Atoms, *Eos Trans. AGU,* **88**(52), Fall Meet. Suppl., Abstract SH12A-0853, 2007.

Khazanov, G.V., and W. Lyatsky, "Relativistic Electrons at Geostationary Orbit: Modeling Results," AGU, Spring Meeting, Ft. Lauderdale, Florida, May 27–30, 2008.

Khazanov, G.V., and W. Lyatsky, Prediction model for relativistic electrons at geostationary orbit, Space Weather Workshop, April 29–May 2, 2008, Boulder, Colorado.

Khazanov, G.V., K. Gamayunov, D.L. Gallagher, and J.U. Kozyra, "Global Simulation of electromagnetic ion cyclotron waves," Fall AGU Meeting, San Francisco, California, December 10–14, 2007 (invited).

Kirk, M.S., and W.D. **Pesnell**, "Methods of detecting polar coronal holes in the EUV," *Eos Trans. AGU,* **88**, Fall Meet. Suppl., Abstract SH13A-1099, 2007.

Klimas, A., and V. Uritsky, "Modeling the turbulent reconnection dynamics of Earth's magnetotail plasma sheet," Fall AGU meeting, 2007.

Klimchuk, J.A., "The angry Sun: Explosions in the corona," George Mason University, 2008 (invited).

Klimchuk, J.A., "Highly efficient modeling of dynamic coronal loops: the EBTEL Hydro Code," International Space Science Institute, Bern, Switzerland, 2008.

Klimchuk, J.A., "EIS observations of nanoflare heating," International Space Science Institute, Bern, Switzerland, 2008.

Klimchuk, J.A., "Understanding warm coronal loops," Fall AGU mtg., San Francisco, California, 2007.

Klimchuk, J.A., "Explaining warm coronal loops," SPD/Spring AGU mtg., Ft. Lauderdale, Florida, 2008.

Klimchuk, J.A., "Coronal loop models and those annoying observations," Hinode II meeting, Boulder, Colorado, 2008 (invited keynote talk).

Kutepov, A. **Feofilov**, L. Rezac, and M. Smith, "Temperatures of Martian atmosphere in the altitude region 60–100 km retrieved from the MGS/TES bolometer infrared limb radiances," Third International Workshop on The Mars Atmosphere: Modeling and Observations, Williamsburg, Virginia, November 10–13, 2008.

Kutepov, A.A., A.G. **Feofilov**, A.S. Medvedev, A.W.A. Pauldrach, and P. Hartogh, "Additional radiative cooling of the mesopause region due to small-scale temperature fluctuations associated with gravity waves," Vol. 10, EGU2008-A-05689, EGU General Assembly, Vienna, Austria, 2008.

Kutepov, A.A., A. **Feofilov**, V.A. Yankovsky, R.O. Manuilova, W.D. **Pesnell**, and R.A. **Goldberg**, "Self-consistent non-LTE model of infrared molecular emissions and oxygen dayglows in the mesosphere and lower thermosphere," *Eos Trans. AGU,* **88,** Fall Mtg. Suppl., Abstract SA41A-0293, 2007.

Lipatov, A.S., and R. Rankin, "Nonlinear magnetic field line resonances, effect of hall term on plasma compression: 1D Hall-MHD Modeling," Planetary and Space Sciences, 2008.

Lyatsky, W., and G.V. **Khazanov**, "A new polar magnetic index of geomagnetic activity and its application to monitoring ionospheric parameters," AGU, Spring Meeting, Ft. Lauderdale, Florida, May 27–30, 2008.

Lyatsky, W., and G.V. **Khazanov**, "Monitoring and forecasting space weather in geospace environment," Space Weather Workshop, Boulder, Colorado, April 29–May 2, 2008, B.

Lyatsky, W., and G.V. **Khazanov**, "A semi-empirical model for forecasting relativistic electrons at geostationary orbit," Fifth Space Weather Symposium, New Orleans, Louisiana, January 20–24, 2008.

MacNeice, P., S. Bakshi, D. Berrios, A. Chulaki, M. Goldfarb, M. **Hesse**, M. Kuznetsova, M. Maddox, K. Patel, A. **Pulkkinen**, L. **Rastaetter**, and A. Taktakishvili, "Heliophysical Modeling at the CCMC—Community Modeling Activities to Compliment the VHGO," paper, Fall AGU 2007, San Francisco, California, December 10–14, 2007.

McGuire, R., D. **Bilitza**, R. Candey, R. Chimiak, J. **Cooper**, S. Fung, D. Han, B. Harris, R. Johnson, J. King, T. Kovalick, H. Leckner, N. Papitashvili, and A. Roberts, "Scientific uses and directions of SPDF data services," *Eos Trans. AGU,* **88**(52), Fall Meet. Suppl., Abstract SH51A-0258, 2007.

Merka, J., T. **Narock**, and A. **Szabo**, "2008 VMO/G status report," HDMC Meeting, Baltimore, Maryland, June 10–12, 2008.

Merka, J., T. **Narock**, and A. **Szabo**, "Relational data searching for magnetospheric data using the Virtual Magnetosheric Observatory," *Eos Trans. AGU,* **88**(52), Fall Meet. Suppl., Abstract SH51A-0247, 2007.

Merka, J., R.J. Walker, T.A. King, and T.W. **Narock**, "Behind the scenes look at the Virtual Magnetospheric Observatory." First Joint Cluster–THEMIS Science Workshop,University of New Hampshire, Durham, New Hampshire, September 23–26, 2008.

Merka, J., A. **Szabo**, R. Walker, T. **Narock**, and T. King, "Uniform data discovery and access with the Virtual Heliospheric and Magnetospheric observatories," EGU General Assembly 2008, Vienna, Austria, EGU2008-A-10989, April 13–18, 2008 (solicited).

Narock, T., T. King, and J. **Merka**, "SPASE Query Language," HDMC Meeting, Baltimore, Maryland, June 10–12, 2008.

Narock, T., A. **Szabo,** and J. **Merka**, "Opportunities for semantic technologies in the NASA heliophysics data environment," AAAI/SSKI08 workshop, talk, 2008.

Pariat, E., "The build-up of current sheets in complex topologies by photospheric driving," American Geophysical Union spring meeting, San Francisco, California, May 2008.

Pariat, E., "Rôle du forçage photosphérique à grande échelle pour la reconnexion magnétique et l'activité solaire," Atelier PNST, Obernai, France, March 2008.

Pesnell, W.D., "The Solar Dynamics Observatory and the wait for solar cycle 24," Solar Image Processing Workshop, Baltimore, Maryland, October 2008.

Pesnell, W.D., "The Solar Dynamics Observatory and the wait for solar cycle 24," Colloquium at the Center for Space and Engineering Research, Virginia Tech, Blacksburg, Virginia, October 2008.

Pesnell, W.D., "What is solar minimum and why do we care?" talk, Heliophysics Subcommittee meeting, NASA Headquarters, Washington, D.C., September 2008.

Pesnell, W.D., "The Solar Dynamics Observatory," talk for Goddard Summer Interns, NASA/Goddard Space Flight Center, Greenbelt, Maryland, August 2008.

Pesnell, W.D., "The Solar Dynamics Observatory: Your eye on the Sun," 37th COSPAR Scientific Assembly in Montreal, Canada, , p. 248, PSW1-0007-08, July 13–19, 2008 (invited).

Pesnell, W.D., "If we can't predict solar cycle 24, what about solar cycle 34?" *AGU Spring Meeting Abstracts*, B2, 2008, (Invited).

Pesnell, W.D., "The Solar Dynamics Observatory," talk for Heliophysics Summer Interns, NASA/Goddard Space Flight Center, Greenbelt, Maryland, June 2008.

Pesnell, W.D., The Solar Dynamics Observatory, SVECSE Workshop, Bozeman, Montana, June 2008

Pesnell, W.D., "Welcome and Opening Remarks," SVECSE Workshop, Bozeman, Montana, June 2008.

Pesnell, W.D., "The Solar Dynamics Observatory," talk at Space Weather Workshop, Boulder, Colorado, April 2008.

Pesnell, W.D., "The Solar Dynamics Observatory," Physics Colloquium, University of Delaware, Newark, Delaware, February 2008.

Pesnell, W.D., "Predicting solar activity: Today, tomorrow, next year," AMS 88th Annual Meeting, 107, 2008 (invited).

Pesnell, W.D., "Variability of solar irradiances using wavelet analysis," *Eos Trans. AGU,* **88,** Fall Meet. Suppl., Abstract SH13A-1100, 2007.

Pesnell, W.D., "The Solar Dynamics Observatory and TIMED/SEE," talk, Research to Operations Workshop, NASA GSFC, Greenbelt, Maryland, November 2007.

Pesnell, W.D., The Solar Dynamics Observatory, Ka-band Ribbon Cutting Ceremony, NASA White Sands Complex, Las Cruces, NM, November 2007.

Pulkkinen, A., R. Pirjola, and A. Viljanen, "Statistical coupling between solar wind conditions and extreme geomagnetically induced current events," paper, Fall AGU 2007, San Francisco, 10–14 December, 2007.

Pulkkinen, A., M. **Hesse**, S. Habib, F. Policelli, M. Kuznetsova, L. **Raestatter**, B. Damsky, L. Van der Zel, D. Fugate, and W. Jacobs, "NASA-EPRI study," paper, EPRI-SUNBURST 2007, Webcast, November 12, 2007 (invited).

Pulkkinen, A., M. **Hesse**, S. Habib, F. Policelli, B. Damsky, L. Van der Zel, D. Fugate, W. Jacobs, "Solar Shield—Forecasting and mitigating solar effects on power transmissions systems," paper, NOAA Space Weather Workshop, April 29–May 2, 2008 (invited).

Pulkkinen, A., "Solar Shield—Forecasting and mitigating solar effects on power transmission systems," presentation, Physics Department, University of Maryland, Baltimore County, Baltimore, Maryland, October 1, 2008 (invited).

Rabin, D.M., R.J. **Thomas**, J.W. **Brosius**, "EUNIS-07: First Look," *Eos Trans. AGU,* **89**, Spring Meet. Suppl., Abstract SP51A-07 (2008).

Rastaetter, L., et al., "Simulations of THEMIS February and March 2008 magnetotail events using global MHD codes at CCMC," 2008.

Rastaetter, L., "CCMC study of dynamic responses to IMF turnings," Univ. of Michigan Cross-Polarcap Potential Workshop, May 21–23, 2008.

Rastaetter, L., "CCMC experiences with OpenGGCM" Univ. of New Hampshire OpenGGCM workshop, October 9–12, 2007.

St. Cyr, O.C., and H. **Xie**, "Solar cycle update (where are the CMEs?)," STEREO SWG #18, April 20–22, 2008.

Saba, J.L.R., and K.T. **Strong**, "Do bursts of activity define the solar cycle? (Do they have forecast potential?)," poster, 5th Space Weather Symposium at the 88th Annual Meeting of the AMS, New Orleans, Louisiana, January 20–24, 2008.

Saba, J.L.R., and K.T. **Strong**, "A new magnetic tool for solar cycle forecasting," poster, Space Weather Workshop, Boulder, Colorado, April 29–May 2, 2008.

Saba, J.L.R., K.T. **Strong**, J.M. **Davila**, and J.R. Herman, "Science Enabled By A High Altitude Airship (HAA)," poster, NASA/Heliophysics Town Hall Meeting—Planning Our Strategy for the Future—College Park, Maryland, May 19–20, 2008.

Šafránková, J., Z. Němeček, A. A. Samsonov, J. **Merka**, and A. **Koval**, "Reaction of the bow shock and magnetopause on interplanetary shocks: Multipoint observations and MHD modeling," EGU General Assembly 2008, Vienna, Austria, EGU2008-A-08960, April 13–18, 2008.

Sahraoui, F., "Fourier phases and intermittency in space plasma turbulence: the technique of Surrogate data," at the 15th Cluster Workshop, Tenerife, March 7–15, 2008 (invited).

Sahraoui, F., and M. **Goldstein**, "Diagnosis of magnetic structures and intermittency in space plasma turbulence using the method of Surrogate data," First Joint Meeting THEMIS-Cluster, Durham, New Hampshire, September 23–26, 2008.

Sahraoui, F., "Recent results on Fourier phases and intermittency in magnetosheath turbulence using the method of surrogate data," seminar, Space Science Laboratory, Berkeley, California, September 30, 2008.

Sahraoui, F., and M. **Goldstein**, "On anisotropies and magnetic structures observed by the Cluster satellites in the magnetosheath/solar wind turbulence," Warwick Turbulence Symposium, November 3–7, 2008, Warwick, United Kingdom (invited).

Schmidlin, F.J., and R.A. **Goldberg**, "Differences and similarities between summer and winter temperatures and winds during MaCWAVE," 37th COSPAR Scientific Assembly, Montreal, Canada, C22-0062-28, July 12–20, 2008.

Schulze-Makuch, D., J. Houtkooper, and J. **Cooper**, "Oxidants: Chemical energy for life on Mars and in the outer solar system," *Eos Trans. AGU,* **88**(52), Fall Meet. Suppl., Abstract P11C-0698, 2007.

Senske, D., L. Prockter, G. Collins, J. **Cooper**, A. Hendrix, K. Hibbitts, M. Kivelson, G. Orton, H. Schubert, A. Showman, E. Turtle, D. Williams, J. Kwok, T. Spilker, and G. Tan-Wang, "The Jupiter System Observer: Probing the foundations of planetary systems, "*Eos Trans. AGU,* **88**(52), Fall Meet. Suppl., Abstract P21B-0531, 2007.

Sittler, E.C., Jr., C. Arridge, A.M. Rymer, A.J. Coates, N. Krupp, M. Blanc, J.D. Richardson, N. Andre, M.F. Thomsen, R.L. Tokar, H.J. McAndrews, M.G. Henderson, J.F. **Cooper**, M.H. Burger, D.G. Simpson, K.K. Khurana, C.T. Russell, M. Dougherty, and D.T. Young, "Cassini observations of Saturn's magnetotail region: Preliminary results," *Eos Trans. AGU,* **88**(52), Fall Meet. Suppl., Abstract SM53A-1088, 2007.

Sittler, E., R. **Hartle**, J. **Cooper**, R. **Johnson**, and A. Coates, "Heavy ion formation in Titan's ionosphere, magnetospheric introduction of free oxygen and source of Titan's aerosols," talk PS4.0-1MO3O-006, EGU General Assembly 2008, Vienna, Austria, April 13–18, 2008.

Sittler, E.C., Jr., J.F. **Cooper** (presenter), S.J. Sturner, R.E. **Hartle**, A. **Lipatov**, P. Mahaffy, T.A. Cassidy, R.E. **Johnson**, N. Andre, and M. Blanc, "Plasma composition science for Europa and the Jovian system," Europa-Jupiter International Science Workshop ESA-ESRIN, Frascati, Italy, April 21–22, 2008.

Sittler, E.C. Jr., R. **Hartle**, J. **Cooper**, R. **Johnson**, and A. Coates, "Heavy ion formation in Titan's ionosphere, magnetospheric introduction of free oxygen and source of Titan's aerosol," Titan Surface Science Meeting, Flagstaff, Arizona, August 19–20, 2008.

Sittler, E.C., Jr., R. **Hartle**, J. **Cooper**, R. **Johnson**, and A. Coates, "Heavy ion formation in Titan's ionosphere, magnetospheric introduction of free oxygen and source of Titan's aerosol," EPSC Abstracts, Vol. 3, EPSC2008-A-00300, Münster, Germany, September 26, 2008.

Strong, K.T., "A personal view of the future of space weather," Space Environment and Commercial Business Activities: Roundtable Session, at the Space Weather Workshop, Boulder, Colorado, April 29–May 2, 2008, (invited).

Strong, K.T., and J.L.R. **Saba**, "Predicting solar cycles," talk, Joint Meeting of AAS/SPD and AGU, Fort Lauderdale, Florida, May 27–30, 2008.

Strong, K.T., and J.L.R. **Saba**, "A new approach to solar forecasting," talk, in session PSW1 on Preparing for the Next Solar Cycle, 37th COSPAR Scientific Assembly, Montreal, Canada, July 13–20, 2008.

Stubbs, T.J., W.M. Farrell, J.S. Halekas, M.R. Collier, G.T. Delory, and R.R. Vondrak, "The lunar dust-plasma environment," LSSO Lunar X-ray Observatory team meeting, Greenbelt, Maryland, October 25, 2007.

Stubbs, T.J., W.M. Farrell, J.S. Halekas, M.R. Collier, G.T. Delory, D.A. Glenar, and R.R. Vondrak, "A heliophysical monitoring network for the near-surface lunar plasma and radiation environments," Lunar Planet. Sci. Conf. XXXIX, #2467, March 10–14, 2008.

Stubbs, T.J., "Mapping lunar surface electric fields and characterizing the exospheric dust environment," CRaTER Science Team Meeting, Boston University, May 13–14, 2008.

Stubbs, T.J., W.M. Farrell, G.T. Delory, M.R. Collier, J.S. Halekas, and R.R. Vondrak, "Tackling the lunar dust-plasma environment: Challenges for science and exploration," 39th Division for Planetary Sciences meeting, Orlando, Florida, October 7–12, 2007.

Stubbs, T.J., W.M. Farrell, J.S. Halekas, M.R. Collier, G.T. Delory, and R.R. Vondrak, "Lunar dust-plasma environment: Heliophysics and planetary update," Lunar Airborne and Dust Toxicity Advisory Group Annual Review meeting, Houston, Texas, November 6–7, 2007.

Stubbs, T.J., W.M. Farrell, M.R. Collier, J.S. Halekas, G.T. Delory, M.P. Holland, and R.R. Vondrak, "Supercharging of the lunar surface by solar wind halo electrons," Fall AGU meeting, San Francisco, California, December 10–14, 2007.

Stubbs, T.J., D.A. Glenar, J.M. Hahn, B.L. Cooper, W.M. Farrell, and R.R. Vondrak, "Predictions for the lunar horizon glow observed by the Lunar Reconnaissance Orbiter Camera," Lunar Planet. Sci. Conf. XXXIX, #2378, March 10–14, 2008.

Stubbs, T.J., "Mapping lunar surface electric fields and characterizing the exospheric dust environment," 5th LRO Project Science Working Group meeting, NASA Goddard Space Flight Center, Greenbelt, Maryland, June 24–26, 2008.

Szabo, A., T. **Narock**, J. **Merka**, A. **Roberts**, J. Vandegriff, G. Ho, J. Raines, P. Schroeder, A. Davis, and J. Kasper, "The Virtual Heliospheric Observatory (VHO)," *Eos Trans. AGU,* **88**(52), Fall Meet. Suppl., Abstract SH51-0250, 2007.

Thomas, R.J., T. Wang, D.M. Rabin, D.B. Jess, J.W. Brosius, "EUNIS Underflight Calibrations of CDS, EIT, TRACE, EIS, and EUVI," NAM-2008 Abstract P37/488, 2008.

Thomas, R.J., T. Wang, D.M. Rabin, D.B. Jess, J.W. **Brosius**, "EUNIS Underflight Calibrations of CDS, EIT, TRACE, EIS, and EUVI," *Eos Trans. AGU,* **89**, Spring Meet. Suppl., Abstract SP51B-04, 2008.

Thompson, W.T., "3D reconstruction of an erupting prominence," Solar Image Processing Workshop IV, October 2008.

Thompson, W.T, and N. **Reginald**, "The radiometric and pointing calibration of SECCHI COR1 on STEREO," AGU Spring Meeting, May 2008.

Thompson, W.T., "Stereoscopic Observations of a Kreutz Sungrazing Comet by the SECCHI COR1 Coronagraphs Aboard STEREO," AGU Fall Meeting, December 2007.

Usmanov, A.V., W.M. Matthaeus, M.L. **Goldstein**, and B. **Breech**, "Three-dimensional MHD solar wind model with turbulence transport," ASTRONUM 2008 3rd International Conference on Numerical Modeling of Space Plasma Flows, St. John, US Virgin Islands, June 8–13, 2008.

Usmanov, A.V., and M.L. **Goldstein**, "Three-dimensional magnetohydrodynamic modeling of the solar corona and solar wind for a highly-tilted solar dipole," AGU Fall 2007 Meeting, San Francisco, California, December 10–14, 2007.

Vandegriff, J., J. **Merka**, T. **Narock**, and A. **Szabo**, "Data unification and analysis services for VxO's," *Eos Trans. AGU,* **88**(52), Fall Meet. Suppl., Abstract SH51A-0257, 2007.

Walker, R.J., J. **Merka**, T.A. King, S.P. Joy, and T. **Narock**, "Access to space weather data through the virtual observatories," SCOSTEP International CAWSES Symposium, Kyoto, Japan, October 23–27, 2007.

Walker, R.J., J. **Merka**, T.A. King, T.W. **Narock**, S. Joy, L. Bargatze, P. Chi, and J. Weygand, "Cluster, THEMIS and the Virtual Magnetospheric Observatory," First Joint Cluster–THEMIS Science Workshop, University of New Hampshire, Durham, New Hampshire, September 23–26, 2008.

Walker, R J., J. **Merka**, T.A. King, T. **Narock**, S.P. Joy, L.F. Bargatze, P. Chi, and J. Weygand, "The Virtual Magnetospheric Observatory." Fifty Years after IGY symposium, Japan, November 2008.

Wang, T., J.W. **Brosius**, D.M. **Rabin**, R.J. **Thomas**, "Radiometric calibration of EUNIS-06 with theoretical predicted 'insensitive' line ratios," *Eos Trans. AGU,* **88**(52), Fall Meet. Suppl., Abstract SH53A-1049, 2007.

Wang, T.J., L. Siu, and J. Qiu, 2008, "Direct observation of high-speed plasma outflows produced by magnetic reconnection in solar impulsive events," Second Hinode Science Meeting, Boulder, Colorado, September 29–October 3, 2008.

Wang, T.J., S.K. Solanki, and M. Selwa, "Identification of different types of kink modes in coronal loops: principles and application to TRACE results," poster, Joint Assembly SPD/AGU, Fort Lauderdale, Florida, May 27–30, 2008.

Wang, T.J., J.W. **Brosius**, R.J. **Thomas**, and D.M. **Rabin**, "Absolute radiometric calibration of EUNIS-06 with theoretically predicted 'insensitive' line ratios," poster, AGU Fall, Solar Physics, San Francisco, California, December 10–14, 2007.

Webb, P.A., UMBC/GEST Heliophysics: Potential for CRESST Collaborations, CRESST Retreat, University of Maryland Baltimore County, Baltimore, Maryland, May 8, 2008.

Webb, P.A., R.F. **Benson**, J.M. Grebowsky, D. **Benson**, V. Truhlik, and X. Huang, "Modeling studies of the topside ionosphere based on ISIS Satellite sounder data," Space Science Division Seminar, Naval Research Laboratory, Washington, DC, June 26, 2008.

Webb, P.A., M.M. Kuznetsova, M. **Hesse**, L. **Benson**, and A. Chulaki, "Ionosphere-thermosphere models at the Community Coordinated Modeling Center," Ionospheric Effects Symposium 2008, Alexandria, Virginia, May 13–15, 2008.

Webb, P.A., R.F. **Benson**, R.E. Denton, J. Goldstein, L. Garcia, and B.W. Reinisch, "An inner-magnetospheric electron density database determined from IMAGE/RPI passive dynamic spectra," 2007 AGU Fall Meeting, San Francisco, California., December 10–14, 2007.

Xie, H., O.C. St. Cyr, N. Gopalswamy, and Q. Hu, "Modeling and prediction of the May 13 2005 event," SHINE workshop, 2008.

Zenitani, S., "The role of the Weibel instability at the reconnection jet front in relativistic pair plasma reconnection," US-Japan Workshop on Magnetic Reconnection (MR2008), Okinawa, Japan, March 2008 (invited).

Zenitani, S., "On the dissipation region in relativistic pair plasma reconnection," Japan Geophysical Union Meeting 2008, Chiba, Japan, May 2008.

Zenitani, S., "The role of the Weibel instability at the reconnection jet front in relativistic pair plasma reconnection," Japan Geophysical Union Meeting 2008, Chiba, Japan, May 2008.

Zenitani, S., "Current sheet expanding processes in relativistic pair plasma reconnection," Cambridge conference 2008, Bohn, Germany, August 2008 (invited).

APPENDIX 3: CURRENT HSD MISSIONS

Interstellar Boundary Explorer (IBEX)

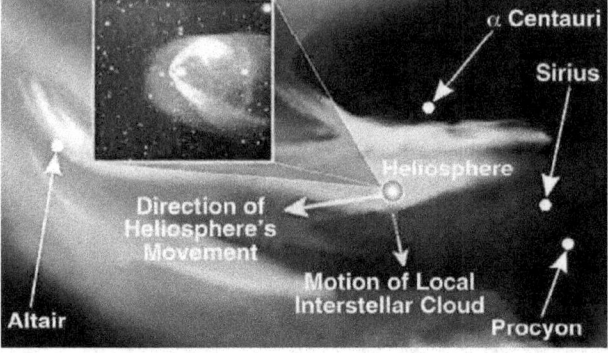

Background: The Sun and solar system move through a part of the galaxy referred to as the local interstellar medium. It is built up from material released from the stars of the Milky Way galaxy through stellar winds, novas, and supernovas. The interstellar medium has considerable structure, as illustrated here. IBEX images reveal global properties of the interstellar boundaries that separate the heliosphere from the local interstellar medium. The Voyager *in situ* measurements are known at one specific location and time. Although Voyager 1 has recorded the first measurement of the termination shock and heliosheath properties, single-point measurements do not give a global understanding of the system. Because IBEX provides global maps of the interstellar interaction, IBEX observations are highly complementary to, and synergistic with, the detailed single-direction measurements provided by the Voyager satellites.

Scientific Goals: IBEX's sole science objective is to discover the global interaction between the solar wind and the interstellar medium. IBEX achieves this objective by taking a set of global energetic neutral atom (ENA) images that answer four fundamental science questions:
- What is the global strength and structure of the termination shock?
- How are energetic protons accelerated at the termination shock?
- What are the global properties of the solar wind flow beyond the termination shock and in the heliotail?
- How does the interstellar flow interact with the heliosphere beyond the heliopause?

Status: IBEX was launched successfully on 2008 October 19

Communications/Navigation Outage Forecasting System (C/NOFS)

Background:
The C/NOFS mission includes a satellite designed to investigate and forecast scintillations in Earth's ionosphere. It was launched on a Pegasus-XL rocket on 2008 April 17 into a low-Earth orbit with an orbital inclination of 13°, a perigee of 400 km, and an apogee of 850 km. The satellite, which is operated by the USAF Space Test Program, will allow the US military to predict the effects of ionospheric activity on signals from communication and navigation satellites, outages of which could potentially cause problems in battlefield situations.

The C/NOFS Spacecraft

C/NOFS includes comprehensive measurements of vector DC and wave electric fields, magnetic fields, plasma density and temperature, ion drifts, neutral winds, lightning detector counts, as well as GPS scintillation and radio beacon experiments. Combined with a strong modeling and ground-based observing component, this mission promises to demonstrably advance scientific understanding of Equatorial Spread F (ESF) irregularities and their conditions for growth.

CINDI Instrument on C/NOFS:
The Coupled Ion Neutral Dynamic Investigation (CINDI) is a Mission of Opportunity investigation on C/NOFS sponsored by NASA and designed and built by the University of Texas at Dallas. CINDI involves two instruments on the C/NOFS satellite that measure the concentration and kinetic energy of the ions and neutral particles in space as the satellite passes through them. This information will be used in building models to understand the various structures in the ionosphere, such as plasma depletions and associated turbulence in the nightside, low-latitude ionosphere. These structures can interfere with radio signals between Earth and spacecraft in orbit, thus causing errors in tracking and loss of communication.

GSFC Role on C/NOFS:
HSD built the Vector Electric Field Instrument (VEFI) for C/NOFS, which consists primarily of an electric field detector that utilizes three orthogonal 20-m tip-to-tip double-probe antennas. VEFI measures direct current (DC) electric fields, which cause the bulk plasma motion that drives the ionospheric plasma to be unstable. Additionally, it measures the quasi-DC electric fields within the plasma density depletions to reveal the motions of the depletions relative to the background ionosphere. VEFI also measures the vector AC electric field, which characterizes the ionospheric disturbances associated with spread-F irregularities. A fluxgate magnetometer, optical lightning detector, and a fixed-bias Langmuir probe are also included in the VEFI instrument package. HSD also manages the MO&DA funding for CINDI, and provides Project Scientist support.

Significant Milestones in FY08:
The C/NOFS mission was successfully launched in 2008. The instruments have been commissioned and are returning excellent data. Among the many initial results revealed by C/NOFS instruments is the fact that the nightside, low-latitude ionosphere is highly structured even during solar minimum and that the majority of spread-F depletions observed thus far occur post-midnight.

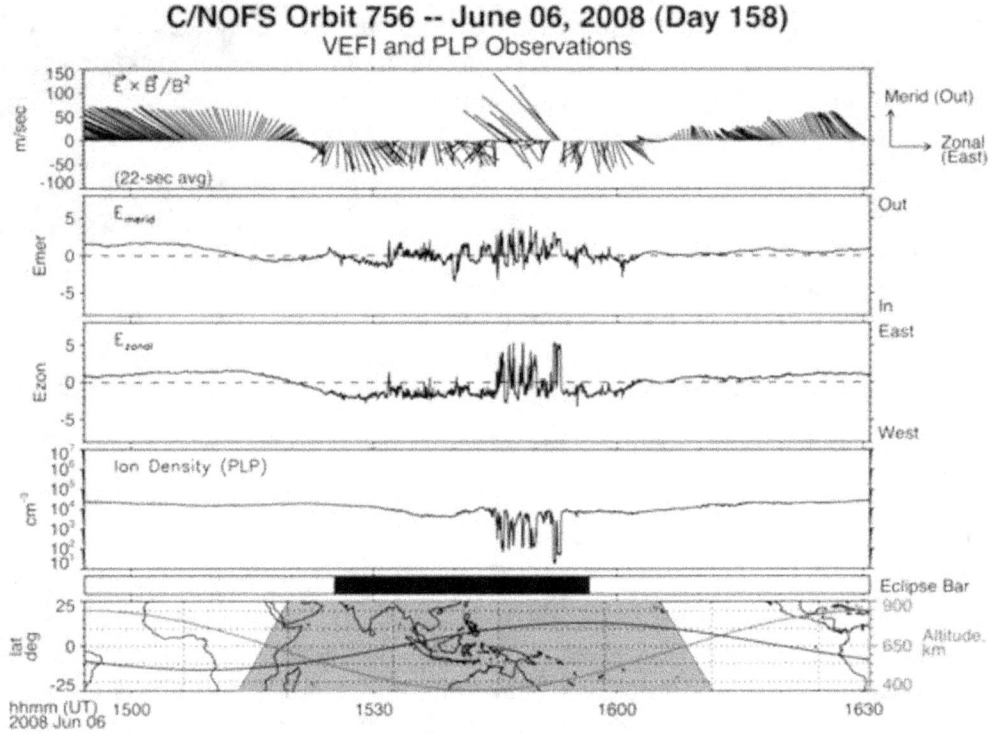

Example of DC electric fields gathered with GSFC's electric field detector.

Aeronomy of Ice in the Mesosphere (AIM)

Noctilucent clouds

Background: The AIM spacecraft was launched from a Pegasus rocket on 2007 April 25 into a 600-km orbit. It has observed two Northern Hemisphere seasons and one Southern Hemisphere season already.

Scientific Goals: AIM is the first satellite mission that probes on a global scale with high spatial resolution, the basic physics of polar mesospheric clouds (PMCs), also known as noctilucent clouds, and makes measurements that can provide information on how these clouds form and vary. AIM addresses the following questions:

o Are there temporal variations in PMCs that can be explained by changes in solar irradiance and particle input?
o What changes in mesospheric properties are responsible for north/south differences in PMC features?
o What controls interannual variability in PMC season duration and latitudinal extent?
o What is the mechanism of teleconnection between winter temperatures and summer hemisphere PMC's?
o What is the global occurrence rate of gravity waves outside the PMC domain?

Despite a significant increase in PMCs research in recent years, relatively little is known about the basic physics of these clouds at "the edge of space" and why they are changing. They have increased in brightness over time, are being seen more often, and appear to be occurring at lower latitudes than ever before. It has been suggested (and debated) that these changes are linked to global climate change.

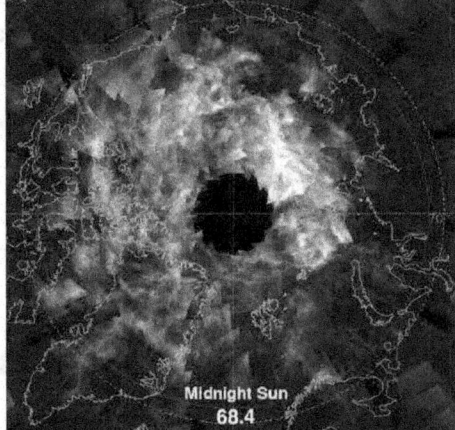

The Cloud Imaging and Particle Size (CIPS) Experiment shows that clouds are highly variable from orbit to orbit and day to day. "Ice voids" were observed that look like tropospheric features.

GSFC Role:
Hans Mayr was the Project Scientist of the AIM mission and has been involved in the scientific analysis.

Significant Project Milestones in FY08:
o The uplink receiver has had several instances of bit lock.
o Science software was released that includes solar source function corrections and adding non-LTE to the temperature retrievals.
o AIM was selected for extended mission funding through 2012, which will allow tracking the evolution of PMCs for an additional seven seasons.

Time History of Events and Macroscale Interactions During Substorms (THEMIS)

Background:

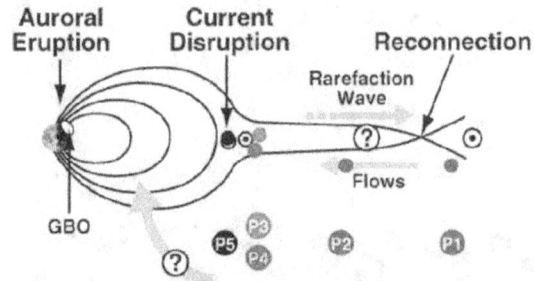

The processes involved in a substorm

THEMIS employs a carefully positioned set of 5 identical spacecraft and 20 science ground observatories located throughout Canada and Alaska, and 12 E/PO ground observatories to pinpoint when and where substorm onset begins—information crucial to determining the cause of geomagnetic substorms. THEMIS was launched on a Delta II rocket on 2007 February 17 into an elliptical orbit. To optimize the science yield of its outermost two probes and to evade detrimental long shadows in March of 2010, HSD proposed to send P1 and P2 into stable, lunar equatorial orbits, where they will form the new mission ARTEMIS. From hundreds of kilometers to 20 R_E separations at lunar distances, probes P1 and P2 will make the first systematic, two-point observations of distant magnetotail phenomena with comprehensive instrumentation.

Scientific Goals:
- Define the onset and evolution of geomagnetic substorms, in particular when, where, and why onset occurs.
- Quantify the processes that generate storm-time MeV electrons in Earth's radiation belts.
- Identify and evaluate the processes that precondition the solar wind prior to its interaction with the magnetosphere.

A collection of ground-based All-Sky Imagers (ASI) captures the aurora brightening caused by a substorm.

GSFC Role:
GSFC's role is to provide the Project Scientist and participate in the scientific analysis.

Significant Project Milestones in FY08:
- Initial results were presented at the Fall AGU in December 2008, and the session was accompanied by a press conference on flux transfer events, hot-flow anomalies, and substorm timing. It was well attended and resulted in numerous television, radio, and newspaper interviews.
- The PI reported an event consistent with the reconnection model; this was given the cover photograph in *Science*. A press teleconference ensued with much attendant publicity.
- THEMIS was well received at the Senior Review: the baseline mission will be continued.
- GRL and JGR special issues were published.

Solar Terrestrial Relations Observatory (STEREO)

Background:
The twin STEREO spacecraft—A and B—were launched on 2006 October 26 from Kennedy Space Center aboard a Delta 7925 launch vehicle. Each spacecraft used close flybys of the Moon to escape into orbits about the Sun near 1 AU; one spacecraft (A) now leads Earth, while the other (B) trails. As viewed from the Sun, the two spacecraft separate at approximately 44–45 degrees per year. Each STEREO spacecraft is equipped with an almost identical set of optical, radio, and *in situ* particles and field instruments provided by US and European investigators.

An un-earthly eclipse of the Sun by the Moon as viewed STEREO B.

Scientific Goals:
The purposes of the STEREO mission are to:
- Understand the causes and mechanisms of CME initiation
- Characterize the CME propagation through the inner heliosphere to Earth
- Discover the mechanisms and sites of energetic-particle acceleration in the low corona and the interplanetary medium
- Develop a 3-D time-dependent model of the magnetic field topology, temperature, density, and velocity structure of the ambient solar wind

GSFC Role:
Goddard, with Swales Aerospace, provided the inner coronagraph, COR1, for the STEREO Sun–Earth Connection Coronal and Heliospheric Investigation (SECCHI). The HSD manages the mission, through the Project Scientist's office and the STEREO Science Center. The Science Center is the focal point for STEREO science coordination as well as for E/PO; it processes the space weather beacon data, archives the STEREO telemetry, mission support data, higher-level instrument data, and analysis software. HSD science team members provide software for SECCHI and engage in science analysis.

Significant Project Milestones in FY08:
- Considerable progress was made in merging images of CMEs from the two spacecraft with modeling efforts to produce 3-D time-dependent scenarios of CME propagation.
- High-time-resolution observations of EUV waves from the two vantage points have brought into question some of the conventional ideas about the cause of these waves.
- Co-rotating interaction regions have been tracked from the Sun out to at least 1 AU.
- In addition to solar results, STEREO has provided many new observations of comets, showing comet tails being disrupted, and sometimes disconnected, by CMEs.
- The nearly continuous wide-angle imagery from the Heliospheric Imager telescopes is being used to look for new asteroids, new variable stars, star quakes, and even extrasolar planets.
- The radio and plasma wave instruments have been able to detect large quantities of impacts against the spacecraft by nanometer-sized particles flowing out from the vicinity of the Sun at near-solar-wind speeds.

Hinode

Background:
Hinode (Japanese for "Sunrise"), formerly known as **Solar-B**, is a mission of the Japan Aerospace Exploration Agency (JAXA) with US, UK, ESA, and Norwegian collaboration. It was launched on an M-V rocket from Uchinoura Space Center, Japan, on 2006 September 22. The satellite maneuvered to the quasi-circular Sun-synchronous orbit over the day/night terminator, which allows near-continuous  observation of the Sun. Hinode is planned as a 3-year mission to explore the magnetic fields of the Sun. It consists of a coordinated set of optical, extreme ultraviolet (EUV), and X-ray instruments (the Solar Optical Telescope, the EUV Imaging Spectrometer, and the X-ray Telescope) to investigate the interaction between the Sun's magnetic field and its corona. Each of the instruments has the highest angular resolution ever achieved in a solar instrument for its spectral band.

Scientific Goals:
Hinode investigates the interaction between the Sun's magnetic field and the corona. The result will be an improved understanding of the mechanisms that power the solar atmosphere and drive solar eruptions. This information will tell how the Sun generates magnetic disturbances and high-energy particle storms that propagate from the Sun to Earth and beyond; in this sense, Hinode will help scientists predict space weather. Hinode is using its three instruments together to unravel basic information to understand:

Hinode has obtained high-resolution images of the poles of the Sun, which show resolved spicules approximately 300 km across.

- How energy generated by magnetic-field changes in the lower solar atmosphere (photosphere) is transmitted to the upper solar atmosphere (corona)
- How that energy influences the dynamics and structure of that upper atmosphere
- How the energy transfer and atmospheric dynamics affects the interplanetary-space environment

GSFC Role:
Three members of the Hinode operations team are based at GSFC and several HSD scientists are involved in the analysis of Hinode data. Sten Odenwald and Ravi Grant perform Hinode E/PO activities at GSFC. The SDAC/VSO at GSFC serves Hinode data to the solar community. The Hinode project is managed by MSFC.

Significant Project Milestones in FY08:
- Hinode's X-band transmitter signal began to experience irregularities in December 2007, and because of increased irregularities in February, the Hinode team is now performing downlink with the backup S-band antenna. Additional JAXA, NASA, and Norwegian ground stations are now providing more downlink opportunities. New compression algorithms are being used to optimize the data downlink via available telemetry.
- Initial results from Hinode were published in a special issue of *Publications of the Astronomical Society of Japan* in November 2007.
- There was a special Hinode issue of *Science* in December 2007.

Solar Radiation and Climate Experiment (SORCE)

Background:
To continue to monitor the Sun's energy output and to cut down on the uncertainty of solar energy input to Earth's climate, NASA launched the SORCE satellite on 2003 January 25. The satellite flies at an altitude of 640 km in a 40° inclination orbit around Earth on a Pegasus rocket. Onboard SORCE are four instruments that will greatly improve the accuracy of the measurements of solar energy. All instruments take readings of the Sun during each of the satellite's 15 daily orbits. The information is transmitted to a ground station at NASA's Wallops Flight Facility in Virginia and a station in Santiago, Chile.

The SORCE satellite

Scientific Goals:
To answer the questions:
- How does the Sun's output vary, and what is the impact on terrestrial climate?
- What aspects of solar variability are influencing the stratospheric ozone layer?

GSFC Role:
Project management (Earth Sciences, Code 610); Doug Rabin (HSD) was Deputy Project Scientist.

Significant Project Milestones in FY08:
- Continued monitoring Total Solar Irradiance through 2008, recording some of the lowest average measurements of solar output since monitoring began in 1979.

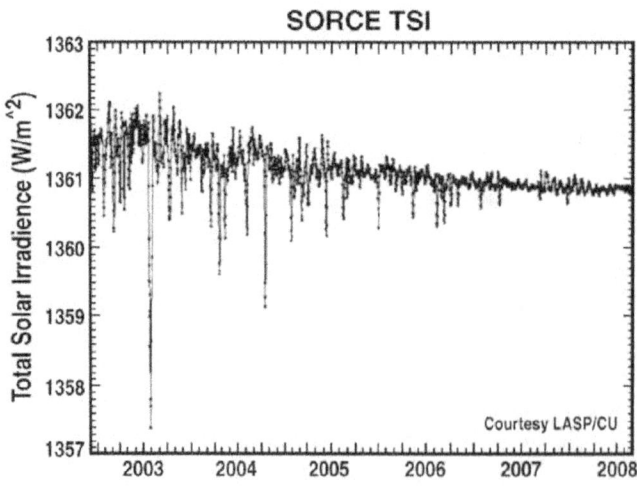

Ramaty High Energy Solar Spectroscopic Imager (RHESSI)

Background:
RHESSI has recorded over 40,000 X-ray flares since its launch on 2002 February 5, and continues to operate successfully. The single RHESSI instrument is an imaging spectrometer observing the Sun in X-rays to γ-rays at time resolutions of a few seconds.

Scientific Goals:
The primary scientific goal of RHESSI is to understand the energy release and particle acceleration during solar flares. This is achieved through X-ray and γ-ray imaging spectroscopy with high angular and energy resolution over the broad energy range from 3 keV to 17 MeV. The focus of the extended mission in FY09 is to integrate new RHESSI flare observations on the rise towards solar maximum with the observations of STEREO, Hinode, SDO, and Fermi (formerly known as the Gamma Ray Large Area Space Telescope, or GLAST), enabling new studies of energy release and particle acceleration processes in flares and CMEs that are more comprehensive than have been previously possible. These include studies of the following topics:

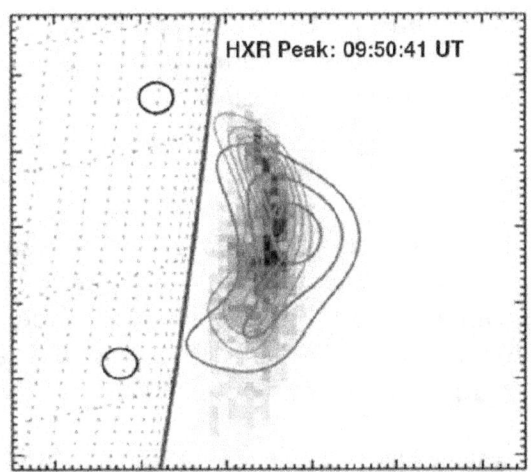

RHESSI and Hinode image of a coronal source at the time of the hard X-ray peak of a flare on 2006 November 21. The thermal source is seen in the Hinode/XRT image at ~2 keV (green) and in the 5–8 keV RHESSI image (red contours). The nonthermal source appears at a slightly higher altitude in the 18–30 keV RHESSI image (blue contours).

- the processes leading up to the flare/CME trigger point
- the initiation of the energy release itself, possibly best revealed by the nonthermal effects even in the weakest microflares
- the location of the electrons and ions in more of the large γ-ray flares
- the location and properties of the coronal hard X-ray sources seen in many flares
- the detailed temporal and spatial comparisons between flares and associated CMEs

GSFC Role:
HSD supplied the flight-qualified cryocooler to maintain the germanium detectors at their operating temperature below 100 K, as well as the flight tungsten grids vital to the Fourier-transform imaging technique. HSD has participated in the mission operations phase through the monitoring of the instrument performance and the scientific analysis of the observations, as well as distributing and archiving the RHESSI data and developing software.

Significant Project Milestones in FY08:
- The 8th RHESSI science workshop was held at the Astrophysical Institute Potsdam (AIP), Germany, on 2008 September 2–6.
- RHESSI received the second highest grade for science merit in the Senior Review.

Thermosphere, Ionosphere, Mesosphere, Energetics, and Dynamics (TIMED)

Background: The mesosphere, lower thermosphere, ionosphere (MLTI) region is a gateway between Earth's environment and space, where the Sun's energy is first deposited into Earth's environment. TIMED is focusing on a portion of this atmospheric region located approximately 60–180 km above the surface. The TIMED spacecraft was launched on 2001 December 7 from Vandenberg Air Force Base, California, aboard a Delta II launch vehicle into a 625 km circular orbit with a 74.1° inclination.

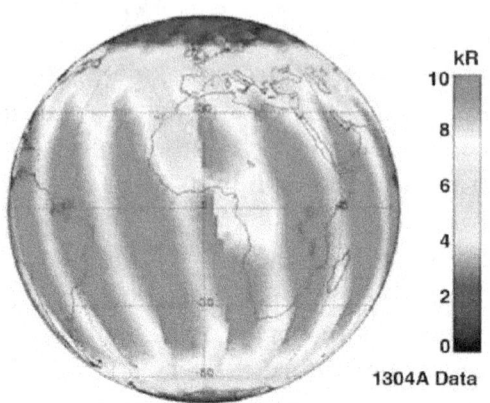

TIMED observes effect of solar eclipse of 2006 March 29 on the thermosphere.

Scientific Goals:
TIMED goals are to characterize the physics, dynamics, energetics, thermal structure, and composition of Earth's MLTI region. The extended mission (2013) objectives are:
- To characterize and understand the solar cycle-induced variability of the MLTI region; and
- To address the processes related to human-induced variability of the mesosphere-lower thermosphere.

GSFC Role:
HSD administered and monitored 17 grants and contracts in 2008, while GSFC has provided oversight of satellite activities. HSD scientists are making fundamental contributions toward the interpretation of SABER data including several presentations and publications.

Significant Project Milestones in FY08:
- A major breakthrough this year was the development of an algorithm for the first extraction of mesospheric water vapor from the Sounding of the Atmosphere using Broadband Emission Radiometry (SABER) data.
- The composite Lyman-α time series has been updated to include the latest versions of the TIMED Solar EUV Experiment (SEE) and SORCE Solar Stellar Irradiance Comparison Measurement (SOLSTICE) solar irradiance measurements of the bright H I 121.5 nm emission.
- TIMED went through the senior review process.

Cluster

Background:
The four-spacecraft Cluster mission was launched in pairs on two Soyuz rockets in July and August 2000. The launches put the four spacecraft into a polar orbit that targeted some of the most important near-Earth regions: solar wind, bow shock, magnetosheath, cusp, magnetopause, plasmapause, and magnetotail. The orbital configuration of Cluster places a nearly regular tetrahedron of four relatively closely spaced observation points in Earth's magnetosphere and solar wind. The tetrahedral configuration provides snapshots of 3-D structures and allows direct measurements of gradients in key plasma parameters. TIMED continues to explore the structure and dynamics of the bow shock and its role in producing energetic ions and electrons, a problem with fundamental scientific applications in space and astrophysics.

Scientific Goals:
The goals of the extended mission include:
- Using the four spacecraft to characterize fluid turbulence in the solar wind
- Investigating the triggering and time evolution of the auroral acceleration by making the first four spacecraft measurements in the auroral acceleration region
- Discovering the role of current disruption in magnetic substorms by making the first four spacecraft measurements in the near-Earth neutral sheet (at 8–10 R_E radial distance); to investigate the role of 3-D electric fields in tail dynamics by repeating the tilting maneuver in the magnetotail
- Studying the cold plasma and wave activity characteristics related to radiation belt high-energy particles in the inner plasmasphere by using low-perigee data
- Studying the magnetosphere at global scales by enhancing common operations

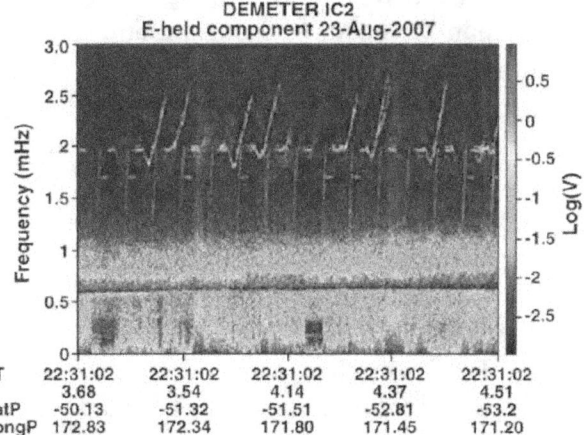

Magnetic shells in space illuminated by powerful ground-based VLF transmitters often contains small-scale plasma density irregularities. VLF waves propagating through these irregularities excite lower-hybrid waves, which then scatter energetic electrons.

GSFC Role:
The two US instrument teams were coordinated through a project office at GSFC. Since the launch, the data analysis grants, and much of the science associated with the US Cluster Co-Is, are managed by the GSFC Cluster Project Scientist. HSD scientists are involved in the analysis of the TIMED data, particularly with Fluxgate Magnetometer (FGM) and Plasma Electron and Current Experiment (PEACE) data.

Significant Project Milestones:
- The first experimental assessment of a physical mechanism that makes a significant contribution to the acceleration of oxygen ions towards the center of the terrestrial magnetic tail, along magnetic field lines.

Two Wide-Angle Imaging Neutral-Atom Spectrometers (TWINS)

Background: TWINS stereoscopically images the magnetosphere and the charge exchange energetic neutral atoms (ENAs) over a broad energy range (~1–100 keV) by using two identical instruments on two widely spaced high-altitude, high-inclination spacecraft. TWINS was launched into two nadir-pointing Molniya orbits at $7.2\,R_E \times 1000$ km at $63.4°$ inclination. Its measurement strategy is to use a neutral-atom imager and the Lyman-α detector, both mounted on a rotating actuator platform to allow 360° azimuthal view. The TWINS instrumentation is essentially the same as the Medium-Energy Neutral Atom (MENA) instrument on the IMAGE mission. This instrumentation consists of a neutral-atom imager covering the ~1–100 keV energy range with $4° \times 4°$ angular resolution, 1-min time resolution, and a simple Lyman-α imager to monitor the geocorona.

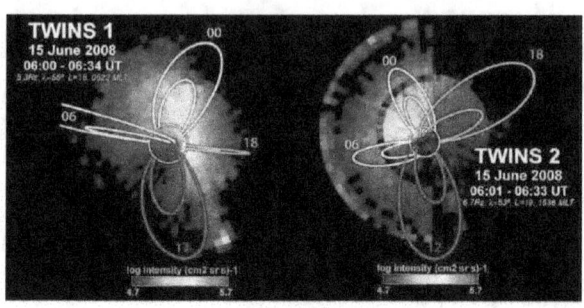

The color indicates ENA flux. Earth is in the center of each image, surrounded by dipole magnetic field lines at 4 and 8 R_E equatorial crossing-points.

Scientific Goals:
- Establish the global connectivities and causal relationships between processes in different regions of Earth's magnetosphere.
- Determine the structure and evolution of the storm-time magnetosphere
- Understand the energization and transport of magnetospheric plasma populations
- Characterize the storm-time sources and sinks of energetic magnetospheric plasma

GSFC Role: The project scientist from HSD participates in weekly science teleconferences, offers advice on achieving TWINS science goals, and assists in interpreting energetic neutral atom (ENA) images. HSD scientists participate in the analysis and interpretation of the TWINS data.

Significant Project Milestones in FY08:
- 2008 April 11: TWINS Flight Model 2 was launched and successfully completed the first-phase checkout.
- 2008 June 15: First-light stereo ENA image obtained by TWINS.
- 2008 November: ENA emissions from low altitude were analyzed and found to have similar energy spectral property to precipitating ions observed by DMSP satellites.

Transition Region And Coronal Explorer (TRACE)

Background: TRACE is a Small Explorer mission designed to study the connections between fine-scale magnetic fields and the associated plasma structures on the Sun in a quantitative way by observing the photosphere, the transition region, and the corona. It was launched by a Pegasus rocket on 1998 April 1 into a Sun-synchronous orbit to get near-continuous observations of the Sun. With TRACE, these temperature domains can be observed nearly simultaneously and at high spatial resolution. TRACE operations will be terminated following intercalibration with the Atmospheric Imaging Assembly (AIA) instrument on the Solar Dynamics Observatory (SDO). SDO is currently manifested for launch in October 2009.

A new-cycle active region (NOAA AR11005) observed by TRACE on 2008 October 14, in Fe XII 195 Å.

Scientific Goals: To simultaneously capture high spatial and temporal resolution images of the transition region. The TRACE data will provide quantitative observational constraints on the models and thus stimulate real advances in understanding the transition region. The data also allow scientists to follow the evolution of magnetic field structures from the solar interior to the corona, investigate the mechanisms of the heating of the outer solar atmosphere, and investigate the triggers and onset of solar flares and mass ejections.

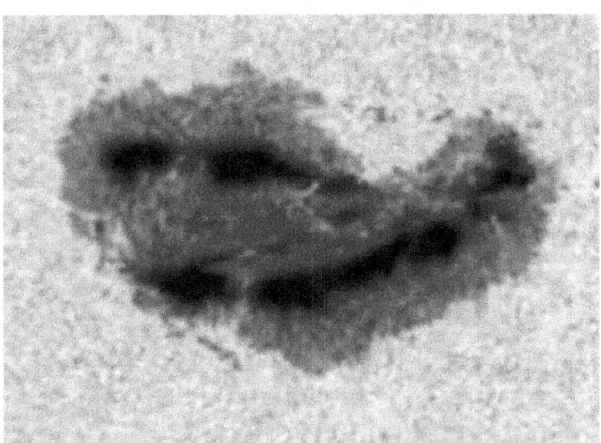

TRACE identified one possible source of the magnetic stress that causes flares: sunspots that rotate, storing energy in the magnetic field.

GSFC Role: GSFC provides mission management (mission scientist and resource analyst), mission operations, scientific operations, and mirrored archive/data access facilities for the TRACE Small Explorer (SMEX) mission. Mission operations are carried out in the SMEX Mission Operations Center (MOC) in Building 3, nearby which is the TRACE Experimenters' Operations Facility (EOF). TRACE data are served by the PI team and via the Virtual Solar Observatory (VSO).

Significant Project Milestones in FY08: A "resident archive" proposal from the PI team (Lockheed Martin) and the Solar Data Analysis Center (SDAC) at Goddard, the first under the new Heliophysics Data Management.

Fast Auroral Snapshot Explorer (FAST)

Background: FAST was launched from Vandenberg Air Force Base onboard a Pegasus XL rocket on 1996 August 21. One in the series of NASA's Small Explorer (SMEX) spacecraft, FAST was designed to observe and measure the plasma physics of the auroral phenomena which occur around both poles of the Earth. It is operated by the University of California Berkeley's Space Sciences Laboratory.

Scientific Goals: Explore the detailed physical processes that accelerate particles and create the visible aurora. Extended mission studies include global aspects of magnetosphere-ionosphere coupling.

FAST Spacecraft

FAST observations provide new understanding of the development and structure of magnetic storms and the ionospheric outflows that provide the plasma that feeds these storms. FAST is the only satellite providing high-time-resolution field and particle measurements at the focus of the magnetic field lines mapping to the magnetopause, cusp, ring current, radiation belts, and magnetotail reconnection region. FAST is the only satellite monitoring ionospheric outflows, a major source of the magnetospheric plasma.

GSFC Role: HSD supplied the project scientist.

Significant Project Milestones in FY08: The FAST mission is now making key contributions to THEMIS and its ground-based observatories' prime mission science and, in 2009, will be in a near-ideal orbit to support Cluster's northern auroral-zone passes, within ~2000 km of Cluster's perigee. This unique opportunity for auroral studies with Cluster will address long-standing questions on the evolution of auroral acceleration as a function of altitude. As such, the FAST mission is in its best position in years to further contribute to many of NASA's key research goals.

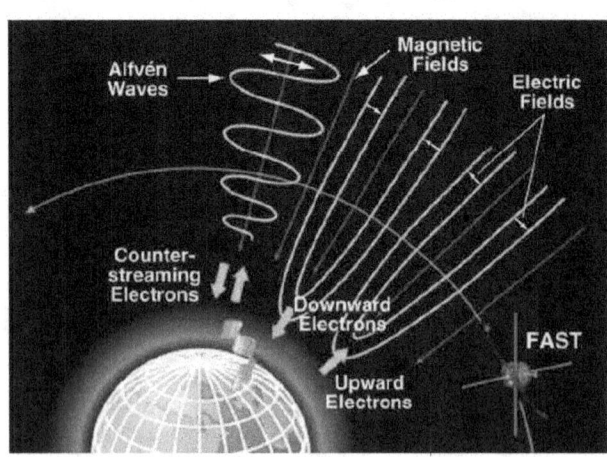

Polar

Background: The Global Geospace Science Polar Satellite was a NASA science spacecraft launched at 06:23:59.997 EST on 1996 February 24 aboard a McDonnell Douglas Delta II 7925-10 rocket from Vandenberg Air Force Base in Lompoc, California, to observe the polar magnetosphere. Polar was designed and manufactured by Lockheed Martin Astro Space Division in East Windsor, New Jersey.

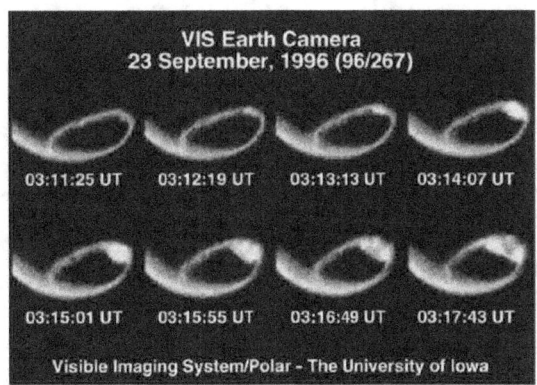

Time sequence of auroral images

Polar is in a highly elliptical, 86° inclination orbit with an orbital period of about 18 hours. It gathers multi-wavelength imaging of the aurora, and measures the entry of plasma into the polar magnetosphere and the geomagnetic tail, the flow of plasma to and from the ionosphere, and the deposition of particle energy in the ionosphere and upper atmosphere.

Scientific Goals: The primary science objectives of the Polar mission are to:
o Measure the mass, momentum, and energy flow and their time variability throughout the solar wind-magnetosphere-ionosphere system that comprises the geospace environment.
o Improve the understanding of plasma processes that control the collective behavior of various components of geospace and trace their cause-and-effect relationships through the system.
o Assess the importance to the terrestrial environment of variations in energy input to the atmosphere caused by geospace plasma processes.

GSFC Role: Two of the twelve PIs for the Polar instruments are located at GSFC, namely for the Thermal Ion Dynamics Experiment (TIDE, T. Moore) and the Visible Imaging System (VIS, J. Sigwarth). An additional two Co-Is are associated with the Electric Field Instrument (EFI, R. Pfaff) and Hydra (K. Ogilivie). The mission is managed from Goddard with J. Sigwarth serving as the project scientist and M. Adrian serving as the deputy project scientist.

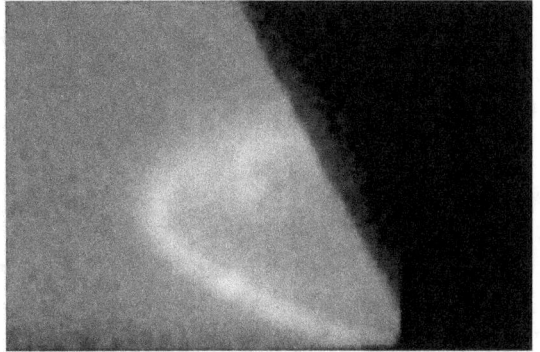

Polar Mission scientists dubbed the mission's final image, taken 16 April 2008, "The Broken Heart" because of its shape.

Significant Project Milestones in FY08:
Polar Mission Operations were terminated at 14:54:41 EDT on 2008 April 28 from the Wind/Polar Mission Operations Room (MOR) in Building 3 at GSFC.

Advanced Composition Explorer (ACE)

Background: ACE was launched in August 1997, carrying six high-resolution instruments designed to measure the elemental, isotopic, and ionic charge-state composition of energetic nuclei from solar-wind to cosmic-ray energies, and three instruments to provide the interplanetary context for these studies. Since January 1998, ACE has been in orbit about the L1 point, 1.5 million km sunward of Earth. Data from ACE are used to study the acceleration and transport of solar, interplanetary, and galactic particles with unprecedented precision.

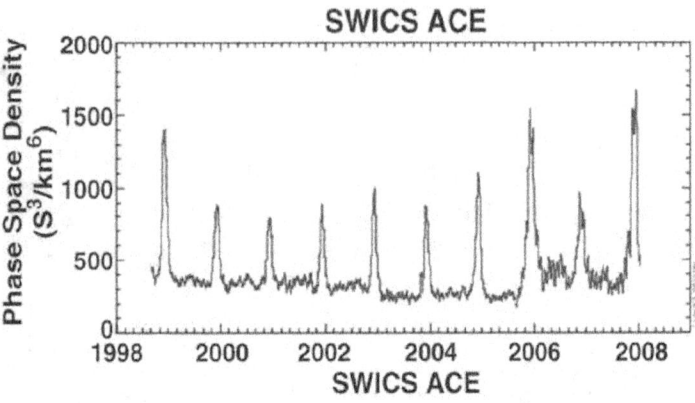

Scientific Goals: The prime objective of ACE is to measure and compare the composition of several samples of matter, including the solar corona, the solar wind, and other interplanetary particle populations, the local interstellar medium, and galactic matter. The scientific questions for the extended mission are:
- How do the compositions of the Sun, solar wind, solar particles, interstellar medium, and cosmic rays differ, and why?
- How does the solar wind originate and evolve through the solar system?
- What is the structure of CMEs and other transients, and how do they evolve?
- How are seed particles fractionated and selected for acceleration to high energies?
- How are particles accelerated at the Sun, in the heliosphere, and in the galaxy?
- How are energetic particles transported in the heliosphere and the galaxy?
- What causes the solar wind, energetic particles, and cosmic rays to vary over the solar cycle?
- How does the solar wind control the dynamic heliosphere?
- How does the heliosphere interact with the interstellar medium?
- How do solar wind, energetic particles, and cosmic rays contribute to space weather over the solar cycle?
- What solar and interplanetary signatures can be used to predict space weather?

GSFC Role: Mission operations support for ACE is provided by the Space Science Mission Operations (SSMO) Project Office, Code 444. The ACE Project Scientist support comes from HSD, and financial administration for grants is provided by HSD. HSD scientists are Co-Is for the Solar Isotope Spectrometer (SIS) and for the Cosmic-Ray Isotope Spectrometer (CRIS).

Significant Project Milestones: The objectives of the ACE Mission for FY09 through FY13 were summarized by the ACE Science Working Team in a proposal for a NASA Headquarters Senior Review in the Spring of 2008.

Solar and Heliospheric Observatory (SOHO)

Background: SOHO has a white-light coronagraph that provides a Sun–Earth line view of both the evolution of, and transient events in, the solar corona; helioseismology and EUV imaging instruments provide baseline intercalibration with SDO analogs before SOHO's end of life in order to extend measurements to a complete, 22-year solar magnetic cycle; continued monitoring of the H I Lyman-α resonant scattering corona, solar wind, and solar energetic particles.

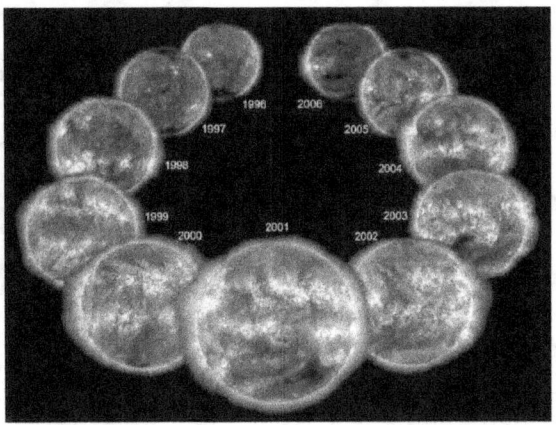

11 years of EUV observations of the Sun

Scientific Goals: There are a number of goals associated with SOHO: in conjunction with SDO and STEREO, understand the causes and mechanisms of CME initiation, and the propagation of CMEs through the heliosphere; continue to monitor the Total Solar Irradiance; monitor the H I Lyman-α corona in order to improve scientific understanding of solar wind acceleration and the distributions of seed particles accelerated as solar energetic particles; continue the measurement of interstellar winds; continue the search for global solar g-modes; and provide operational predictions of solar energetic particles during manned space missions.

GSFC Role: GSFC provides project management, mission operations, scientific operations, an analysis facility, and archive/data access facilities for the SOHO mission. The Solar Data Analysis Center (SDAC) houses, among other data sets, all SOHO data other than the Michelson Doppler Imager (MDI) helioseismology archive and serves it to the worldwide scientific community via the Internet, through both the SOHO archive interface and the Virtual Solar Observatory (VSO). Unfortunately, SOHO has been in operation long enough (13 years) that there is no funding for science in the project budget.

Significant Project Milestones in FY08:
- 1500th comet discovery
- A strong correlation has been identified between enhancements in GOES X-ray flux and total power in 5-min band global *p*-mode oscillations observed in integrated intensity by the Variability of Solar Irradiance and Gravity Oscillations (VIRGO).
- SOHO has observed the entire evolution of solar cycle 23.
- As a result of the Senior Review of Heliophysics Missions held earlier this year, SOHO will continue to operate at least through FY12.
- The SOHO mission was ranked with the Hubble Space Telescope (HST) and the X-ray Multi-mirror Mission (XMM)/Newton as the highest scientific priorities of recent senior reviews of solar system and astrophysics missions by the ESA Space Science Advisory Committee (SSAC). The ESA Scientific Programme Committee will meet in February 2009 to decide on the availability of funding.

Wind

Background: Wind is a comprehensive solar wind laboratory for long-term *in situ* solar wind measurements. Wind is a spin-stabilized spacecraft launched in 1994 November 1 and placed in a halo orbit around the L1 Lagrange point, more than 200 R_E upstream of Earth to observe the unperturbed solar wind that is about to impact the magnetosphere of Earth. Wind, together with Geotail, Polar, SOHO, and Cluster, constitute a cooperative scientific satellite project designated the International Solar Terrestrial Physics (ISTP) program, which aims at gaining improved understanding of the physics of solar-terrestrial relations.

Wind provides a third point of solar wind observations enhancing the science return of the STEREO mission as well as continuing to monitor the solar wind input for geospace studies.

Scientific Goals: The primary science objectives of the Wind mission are as follows:
- Provide complete plasma, energetic particle, and magnetic field measurements for magnetospheric and ionospheric studies
- Investigate basic plasma processes occurring in the near-Earth solar wind
- Provide baseline, 1 AU, ecliptic plane observations for inner and outer heliospheric missions

GSFC Role: Three of the still-functioning seven instruments were developed at GSFC, namely the Magnetic Field Investigation (MFI; A. Szabo, PI), the electron analyzer of the Solar Wind Experiment (SWE; K. Ogilvie, PI), and the high-energy particle instrument (Energetic Particles, Acceleration, Composition, and Transport—EPACT; T. Von Rosenvinge, PI). Moreover, a significant portion of the radio and plasma waves instrument was provided by GSFC (M. Kaiser, PI). The mission is also managed from Goddard with A. Szabo serving as the Project Scientist and M. Collier serving as the Deputy Project Scientist.

Significant Project Milestones in FY08:
- First detection of oppositely directed reconnection jets in a solar wind current sheet
- Evidence for extremely long (at least 390 R_E) reconnection X-lines in the solar wind
- Evidence for quasi-steady reconnection in the solar wind
- Evidence for component reconnection in the turbulent solar wind
- Compositional signatures of CMEs and the slow solar wind
- Tracking and triangulating with STEREO the propagation of interplanetary shocks using Type II radio bursts
- Demonstrating significant deformation of magnetic clouds at 1 AU from pure cylindrical symmetry

Geotail

Background: Geotail was launched in 1992 as a US-Japanese joint mission. It crosses all boundaries through which solar wind energy, momentum, and particles must pass to enter the magnetosphere. Knowledge of the physical processes operating at these boundaries is vital to understanding the flow of mass and energy from the Sun to Earth's atmosphere. The long-lived Geotail spacecraft continues to provide critical and unique geospace measurements essential to fulfilling the key objectives of the Heliophysics Great Observatory at minimal cost.

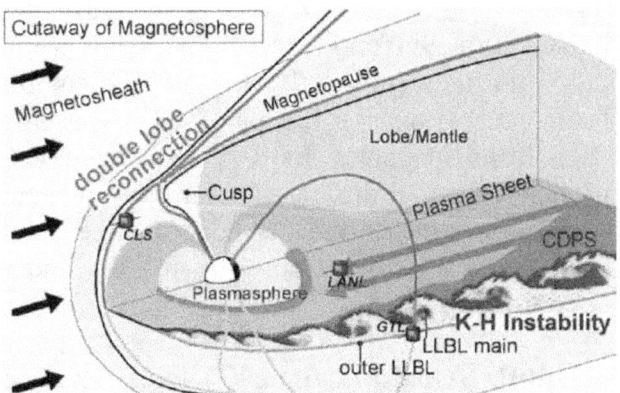

Advances in magnetospheric physics often result from multi-spacecraft coordination as shown here. Geotail made measurements of the Kelvin-Helmholtz instability in the low-latitude dusk boundary layer while Cluster simultaneously detected high-latitude reconnection near the noon meridian.

Scientific Goals: During the current extended mission, the Geotail science focuses are:
o Providing extensive coverage of the magnetospheric boundary layer to delineate mechanisms controlling the entry and transport of plasma into the magnetosphere that is then energized to produce magnetic storms
o Providing supplementary measurements to THEMIS to reveal the spatial and temporal scales of substorm phenomena in the magnetotail
o Providing near-Earth plasma and magnetic field measurements as Geotail spends about 35% of its time in the solar wind
o Providing an important complementary data source for validation of global simulations
o Providing observations that define the location and physics of tail magnetic reconnection and particle acceleration
o Determining energetic particle environments up to, and including, penetrating gamma rays

GSFC Role: GSFC provides ground data system support for Geotail. Deep Space Network (DSN) telemetry data are transferred to the GSFC data system, which performs the initial reduction and merging with trajectory data provided by the Japanese. Data for the Japanese and US experiments are stored and can be accessed by both Japanese and US experimenters. Key parameters are produced directly from the DSN playback data at GSFC and are available from the Coordinated Data Analysis Web (CDAWeb).

Ulysses

Background: The Ulysses spacecraft was launched in 1990 and entered polar orbit around the Sun after a Jupiter flyby in 1992. This joint NASA–ESA mission has provided a wealth of data from magnetic field, plasma, energetic particle, dust, radio, and solar corona radio occultation measurements over three solar orbits. The mission is now nearing the end of operations.

Ulysses explores the heliosphere.

Scientific Goals: Original Ulysses mission science objectives were to investigate, as a function of heliographic latitude, the properties of the solar wind, the structure of the Sun/wind interface, the heliospheric magnetic field, solar radio bursts and plasma waves, solar X-rays, solar and galactic cosmic rays, and both interstellar and interplanetary neutral gas and dust. Additional objectives include the study of the Jovian magnetosphere during the Jupiter flyby, the detection of cosmic γ-ray bursts and triangulation on burst locations with other detectors, and the search for gravitational waves. New objectives have been evolving since launch. These entail combining Ulysses *in situ* measurements of solar-wind fields and particles, cosmic rays, and radio waves over a wide range of heliolatitudes and radial distances with remote observations of the Sun and solar corona from ongoing and upcoming missions. This is being done to analyze properties and dynamics of coronal mass ejections and of sources of the solar wind in order to enhance the ability of those missions to meet their own science objectives and to construct models of the 3-D Sun and heliosphere.

GSFC Role: The HSD Space Physics Data Facility (SPDF) has been the primary NASA distribution point for Ulysses data. HSD members have long participated in science working team meetings for Ulysses and have been in regular contact with Ulysses scientists to define the requirements for, and support the delivery of, experiment and ancillary data products. SPDF supports public access to Ulysses data through highly used data systems such as the Coordinated Data Analysis Web (CDAWeb), the Coordinated Heliospheric Observations Web (COHOWeb), ftp browsing and downloads, and the HelioWeb service providing spacecraft position data in heliocentric coordinates. After many years of less frequently updating SPDF holdings through more occasional contacts with the data providers, SPDF now automatically ingests data updates on a daily basis. SPDF also works closely with the new Virtual Energetic Particle Observatory (VEPO) on virtual observatory support for energetic-particle data products from three instrument suites on Ulysses. The VEPO project is managed by the Chief Scientist for SPDF. Finally, SPDF works with the National Space Science Data Center in the Solar System Exploration Division on delivery and registration of Ulysses data for permanent archiving. SPDF will support the final group of Ulysses resident archives on post-mission delivery of final data products to NSSDC.

Voyager

Background: The Voyager spacecraft continue their epic journey of discovery, traveling through a vast unknown region of the heliosphere on their way to the interstellar medium. Both Voyagers are now traversing the heliosheath, with the first crossings of the heliopause and the first *in situ* observations of the interstellar medium still to come. The twin Voyager 1 and 2 spacecraft continue exploring where nothing from Earth has flown before. Now in the 30th year after their 1977 launches, they each are much farther away from Earth and the Sun than Pluto is and are approaching the boundary region—the heliopause—where the Sun's dominance of the environment ends and interstellar space begins. Voyager 1, more than three times the distance of Pluto, is farther from Earth than any other human-made object and speeding outward at more than 17 km/s. Both spacecraft are still sending scientific information about their surroundings through the DSN.

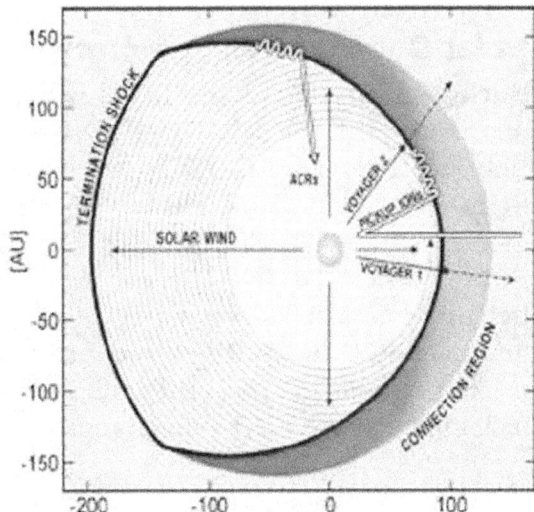

Schematic equatorial view of the termination shock. The color indicates field intensity increasing with time as the field line connection points move to the flanks. Anomalous cosmic rays diffuse along the field line toward the nose of the heliosheath (from McComas and Schwadron, 2006).

Scientific Goals: The goals for the Voyager spacecraft are to explore the interaction of the heliosphere with the local interstellar medium and to study the heliosheath. Major mysteries remain unresolved, such as the source of, and acceleration mechanism for, the anomalous cosmic rays. The Voyager Interstellar Mission (VIM), in combination with IBEX, should be able to solve some of these questions. The nature of the solar wind turbulence and the behavior of major solar wind structures downstream of the termination shock will also be examined by the VIM.

GSFC Role: GSFC's principal contribution is through the Magnetometer (MAG) instrument.

Significant Project Milestones in FY08: The two Voyager spacecraft have been heading out of the solar system for over 30 years. Voyager 2 crossed the termination shock last year at about 8 billion miles from the Sun. An unexpected finding was that the termination shock was re-forming on a time scale of a few hours as Voyager 2 crossed the shock at least five times within three days. The detailed structure of the magnetic field and plasma of the fully developed shock was observed on one crossing. The solar wind speed in the heliosheath behind the shock was still supersonic with respect to the shocked solar wind, which was cooler and faster moving than anticipated. The scales of the magnetic field fluctuations are consistent with the hypothesis that ions derived from the interstellar neutral gas dominate the temperature of the heliosheath.

APPENDIX 4: FUTURE MISSIONS

Solar Dynamics Observatory (SDO)

Background: SDO is the first Living With a Star (LWS) mission. It will use telescopes to study the Sun's magnetic field, the interior of the Sun, and changes in solar activity. Some of the telescopes will take pictures of the Sun, while others will view the Sun as if it were a star.

Scientific Goals: The primary goal of the SDO mission is to understand—driving towards a predictive capability—the solar variations that influence life on Earth and humanity's technological systems by determining:

The SDO spacecraft

o How the Sun's magnetic field is generated and structured, and
o How this stored magnetic energy is converted and released into the heliosphere and geospace in the form of solar wind, energetic particles, and variations in the solar irradiance.

GSFC Role: GSFC built the spacecraft and Dean Pesnell (HSD) is the Project Scientist. Several HSD people are part of the science investigation and help with E/PO.

Status: Early in FY08, SDO was integrated into an observatory by the GSFC engineering team. SDO then entered into a series of environmental tests to assure the team that everything was working. Shake, vibration, electromagnetic interference, and thermal tests were completed by August 2008. The ground system was also completed and tested during FY08. After launch, SDO will be managed by GSFC.

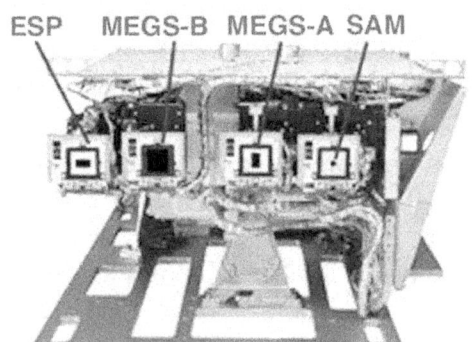

SDO is built, tested, and ready to launch!

The EVE experiment

Radiation Belt Storm Probes (RBSP)

Background: The distances separating the two RBSP spacecraft will vary over the course of the mission, enabling researchers to distinguish between spatial and temporal effects, and identify the spatial extent of various phenomena.

Scientific Goals: The mission goals are:
- Identify and quantify the cause of radiation belt electron enhancement events
- Identify and quantify the dominant mechanisms for relativistic electron loss
- Determine how the ring current and other geophysical phenomena affect the radiation belts

GSFC Role: GSFC retains overall technical authority for the mission, but applies "light-touch" management over this mission, which has been assigned to JHU/APL. In addition, all Project Science responsibilities have been assigned to JHU/APL. GSFC retains the mission science component, which includes review of project science activities.

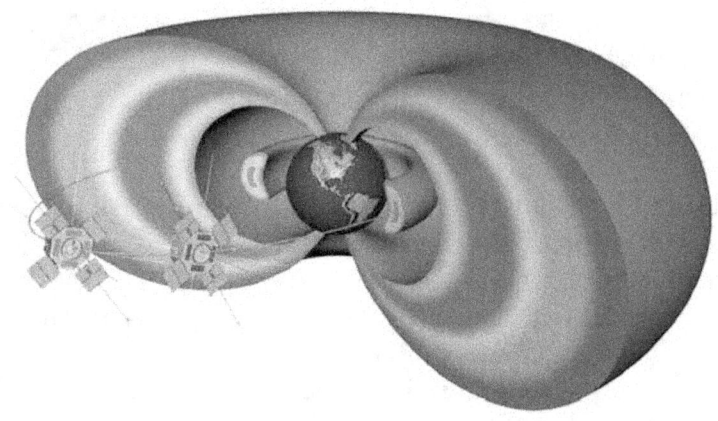

Status:

The mission passed MDR in October 2007 and finalized its Level 1 requirements by September 2008.

Magnetospheric MultiScale (MMS)

Background: The details of how nature releases large amounts of energy are not well understood, although the conversion of magnetic energy from reconnection of oppositely directed fields into heated plasma and energetic particles appears to play a central role. It has not been possible to replicate the typical conditions in the laboratory, nor to diagnose the physical processes that control the release of energy, particularly at the small spatial scales where the electrons become unmagnetized.

Magnetic reconnection at the magnetospheric boundary

Scientific Goals: Observations of reconnection require the three-dimensional determination of the magnetic field and plasma in the vicinity of the site of particle energization. Thus, the overall goal of the MMS mission is to understand the microphysics of magnetic reconnection by determining the kinetic processes responsible for the initiation and evolution of magnetic reconnection. More specifically, MMS will determine the role played by electron inertial effects and turbulent dissipation in driving magnetic reconnection. MMS will also determine the rate of magnetic reconnection and the parameters that control it.

GSFC Role: MMS is a major GSFC product and involves many resources in HSD. Overall, the project is managed at GSFC and the four spacecraft will be built at GSFC. The Project Scientist, Melvyn Goldstein, and the two Deputy Project Scientists, Mark Adrian and Guan Le, are all members of HSD, as is the Lead Investigator for the Fast Plasma Instrument (FPI), Tom Moore. In addition, the Dual Electron Sensors will be built and tested at GSFC; the Japanese Dual Ion Sensors are the responsibility of HSD, and HSD scientists are members of the Fields and the Theory and Modeling team.

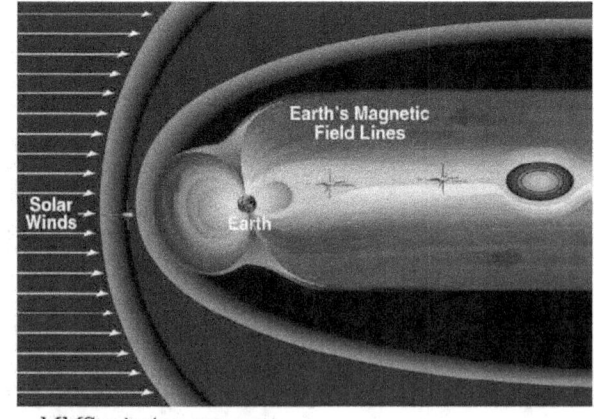

MMS mission concept

Status: The primary goal of MMS in FY08 was to obtain permission to proceed to Phase C/D and to start building hardware. That process will involve many reviews of instruments and of the project during 2009. Locally, as part of that process, a major renovation and expansion of the HSD laboratory facilities is now under way and the Fast Plasma Instrument team is preparing for the reviews, which begin in December 2008.

Solar Orbiter

Background: The Sun's atmosphere and the heliosphere are unique regions of space, where fundamental physical processes common to solar, astrophysical, and laboratory plasmas can be studied in detail and under conditions impossible to reproduce on Earth. The results from SOHO and Ulysses have enormously advanced scientific understanding of the solar corona, the associated solar wind, and the 3-D heliosphere. However, the point has been reached where *in situ* measurements much closer into the Sun, combined with high-resolution imaging and spectroscopy at high latitudes, promise to bring about major breakthroughs.

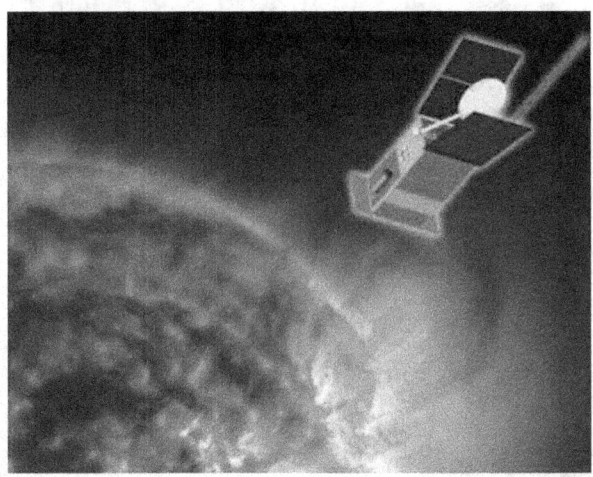

Scientific Goals: The broad science objective is to gain a better understanding of solar activity and variability. In practice, it means extending scientific knowledge of the solar interior to higher latitudes and greater depths, beyond what SDO can achieve. Studying the near-surface layers at high latitudes is a prime objective. Probing the deep interior with sufficient accuracy may, however, require longer observation durations than will be possible. In addition, demonstrating the concept of stereoscopic helioseismology is an important objective. SDO and/or ground-based facilities, together with Solar Orbiter would make a most powerful combination.

GSFC Role: GSFC will host the US Solar Orbiter Project Office following any selection of instruments for that payload, which has not been formally selected as of this writing.

Status: The Solar Orbiter mission was approved in October 2000 by ESA's Science Programme Committee as a flexi-mission for launch in the 2008–2013 time frame.

Solar Probe Plus

Background: Solar Probe Plus (SP+) is humanity's first visit to the Sun to explore the complex and time-varying interplay of the Sun and Earth, which affects human activity. SP+ will determine where and what physical processes heat the corona and accelerate the solar wind to its supersonic velocity. A combined remote-sensing and *in situ* sampling from within the solar corona itself will provide a "ground truth" never before available from astronomical measurements made from spacecraft in Earth's orbit or LaGrange points. SP+ is currently under study as part of NASA's SMD.

The baseline mission provides for 24 perihelion passes inside 0.16 AU (35 R_S), with 19 passes occurring within 20 R_S of the Sun. The first near-Sun pass occurs 3 months after launch, at a heliocentric distance of 35 R_S. Over the next several years, successive Venus gravity assist (VGA) maneuvers gradually lower the perihelia to ~9.5 R_S—by far the closest any spacecraft has ever come to the Sun. The spacecraft completes its nominal mission with three passes, separated by 88 days, at this distance.

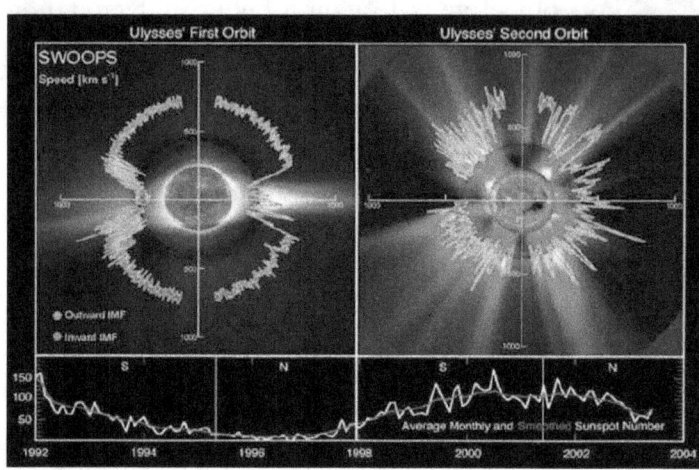

Solar wind speed as measured by Ulysses.

Scientific Goals: Although the SP+ science objectives remain the same as those established for Solar Probe 2005, the new mission design differs dramatically from the 2005 design (as well as from all previous Solar Probe mission designs since the 1970s). The 2005, and earlier, missions involved one or two flybys of the Sun at a perihelion distance of 4 R_S by a spacecraft placed into a solar polar orbit by means of a Jupiter gravity assist. In contrast, SP+ remains nearly in the ecliptic plane and makes many near-Sun passes at increasingly lower perihelia.

GSFC Role: GSFC will provide a mission scientist and will participate in instrument proposals.

Status: A NASA Announcement of Opportunity soliciting instrument proposals is expected in early 2009.

Sentinels

Background: The Solar Sentinels mission is the heliospheric portion of the global Living With a Star (LWS) program connecting the solar and geospace observations made by the Solar Dynamics Observatory (SDO) and the Radiation Belts Storm Probes (RBSP). Sentinels consists of four identical spinning spacecraft launched into slightly different 0.25×0.76 AU near-ecliptic inner heliospheric orbits. The focus of the Sentinels mission is to make multi-point *in situ* observations of the solar wind and the energetic-particle environment.

Sentinels is also the NASA component of the joint NASA–ESA Heliophysical Explorers (HELEX) mission, which also includes ESA's Solar Orbiter mission. Solar Orbiter will focus on out-of-ecliptic and remote-sensing observations of the Sun. The comprehensive inner heliospheric observations will be completed by the upper coronal (10 R_S) *in situ* measurements to be provided by the SP+ mission.

Sentinels will require an Atlas V-class launch vehicle and will use multiple Venus gravity assist maneuvers to attain their orbits.

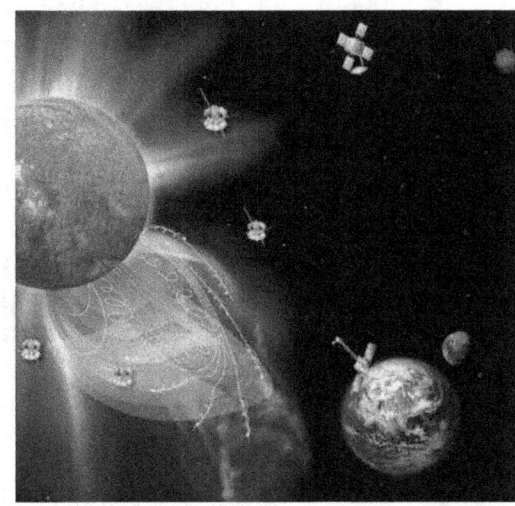

Artist concept of the Sentinels mission from the Sentinels Science and Technology Definition Team (STDT) Report (Courtesy of J. Rumburg).

Scientific Goals: Sentinels has three primary science goals:

1. What are the origins of the solar wind streams and the heliospheric magnetic field?
    a. How and where do fast and slow solar wind streams originate?
    b. What are the solar sources of the heliospheric magnetic field?
    c. What is the solar origin of turbulence and structures at all scales in the solar wind?
2. What are the sources, acceleration mechanisms, and transport processes of solar energetic particles?
    a. What are the sources of energetic particles and how are they accelerated to high energy?
    b. How are solar energetic particles released from their sources and distributed in space and time?
3. How do CMEs evolve in the inner heliosphere?
    a. How is the structure of interplanetary CMEs related to their origin?
    b. How do transients add magnetic flux to, and remove it from, the heliosphere?
    c. How and when do shocks form near the Sun?

GSFC Role: Adam Szabo (GSFC/HSD) is the Study Scientist.

Status: Current plans call for a launch date no earlier than 2017.

Solar C

Background: Following the success of Yohkoh (Solar A) and Hinode (Solar B), JAXA is considering developing a new solar mission. Currently, there are two scientific thrusts under consideration.

Scientific Goals: Depending on which concept is finally decided upon, the scientific goals are either:
- The global versus the local dynamo
    - Measure meridional flow at high latitudes, and see where it turns downwards
    - Detect the magneto-sound speed anomaly located in the tachocline region
    - Observe the vector magnetic fields of the photosphere and chromosphere and obtain coronal imaging in X-ray/EUV wavelengths
    - Obtain acoustic speed and angular rotation speed distributions in the polar region
    - Understand the acceleration mechanism of the fast solar wind
    - Monitor total irradiance (optional)
    - Study the influence of the Sun on the heliosphere

Or,
- Coronal and chromospheric heating
    - Obtain precise chromospheric and, if possible, coronal vector magnetic field maps in addition to photospheric magnetic maps with high spatial and temporal resolution
    - Obtain coronal 3-D magnetic-field map from chromospheric vector field, predict location and evolution of neutral sheets, discontinuities for transient and stationary coronal heating and eruption
    - Reveal causal relationship of photosphere-chromosphere-transition region-corona to understand coronal/chromospheric heating and dynamics
    - Understand the nature of hidden magnetism: Is the observed magnetic field the tip of the iceberg?
    - Deepen Hinode discoveries with quantitative analysis: waves, turbulence, magnetic reconnection
    - Study the influence of the Sun on the heliosphere

GSFC Role: TBD

Status: The concept is being developed by a Solar-C Scientific Working Group. NASA has not committed to participating in the mission as yet.

Geospace Electrodynamic Connections (GEC)

Background: The GEC mission is a multi-spacecraft Solar Terrestrial Probe (STP) that has been specifically designed to advance to a new, and deeper level of physical insight, scientific understanding of the coupling among the ionosphere, thermosphere, and magnetosphere. GEC is NASA's fifth STP. The ionosphere-thermosphere (IT) system is not merely a passive absorber of magnetospheric energy; it is an active participant in the energy exchange process. GEC will, therefore, also investigate the role of the ionosphere and thermosphere in modulating the energy exchange with the magnetosphere and will address a second fundamental question: How is the IT region dynamically coupled to the magnetosphere?

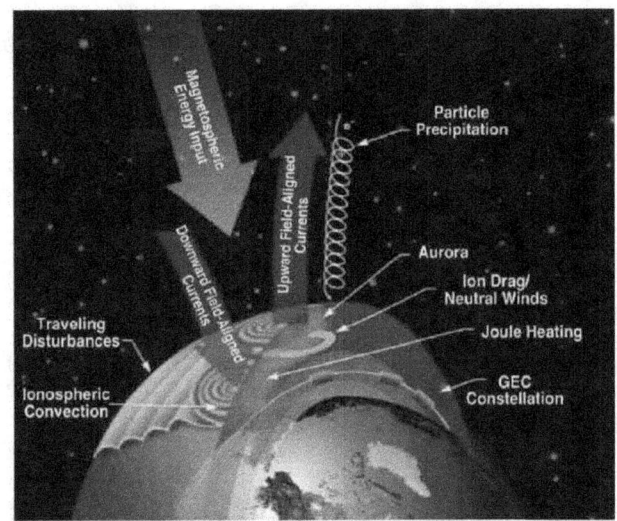

The GEC spacecraft will all be identically instrumented to sample *in situ* the ionized and neutral gases of the upper atmosphere and to measure the electric and magnetic fields that couple the IT system to the magnetosphere. The focus of the GEC mission will be on the lower reaches of the ionosphere and thermosphere, where the neutral atmosphere plays a preeminent role in processing and dissipating the electromagnetic energy received from the magnetosphere. The GEC spacecraft will use onboard propulsion to perform many excursions to altitudes below the nominal perigee of 185 km, sampling a significant portion of the Hall and Pedersen conductivity layer where significant closure of field-aligned currents and associated Joule heating begin to occur. During these low-perigee excursions, and at certain other times as well, GEC *in situ* measurements will be coordinated with ground-based observatories. Such coordinated campaigns are an integral part of the GEC mission concept.

Scientific Goals: Through multipoint measurements in the IT system, GEC will do the following:
- Discover the spatial and temporal scales on which magnetospheric energy input into the IT region occurs
- Determine the spatial and temporal scales for the response of the IT system to this input of energy
- Quantify the altitude dependence of the response. GEC will thereby answer the fundamental question: How does the IT system respond to magnetospheric forcing?

GSFC Role: TBD

Status: Because of recent budget cuts impacting the STP Program, the GEC mission is moved outside the near-term (5-year) budget planning window. Status updates associated with recent roadmap activities are currently under way; however, beyond roadmap activity updates, minimal updates are planned for GEC.

Magnetospheric Constellation (MagCon)

Background: MagCon is a constellation of 50 small satellites distributed in 3 x 7 R_E to 3 × 40 R_E, low-inclination, nested orbits, with "nearest-neighbor" average spacing 1.0–2.0 R_E between satellites, in the domain of the near-Earth plasma sheet.

Scientific Goals: MagCon will answer the fundamental question:

o How does the dynamic magnetotail store, transport, and release matter and energy?

GSFC Role: Build on ST5 technology development

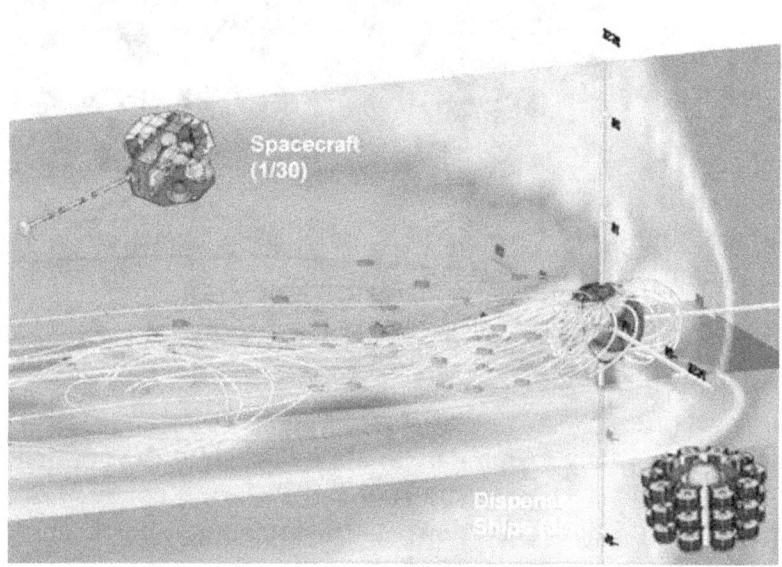

Status: Future funding for the MagCon mission is highly uncertain and has tentatively been deferred beyond FY09 by the President's FY05 budget proposal. NASA HQ, however, has requested a white paper summarizing the status of this mission and the development options for future work, in the event that funding becomes available.

Interstellar Probe

Background: Interstellar Probe will include a comprehensive suite of sensors designed to measure the detailed properties of the plasma, neutral atoms, energetic particles, magnetic fields, cosmic rays, and dust at the heliospheric boundaries and in the nearby interstellar medium. It will explore the "wall" of neutral interstellar hydrogen that lies just beyond the heliopause and determine the large-scale structure of the heliosphere by imaging energetic neutral atoms created in dynamic processes occurring just beyond the termination shock. In addition, Interstellar Probe will map the infrared emission in the outer solar system to determine the distribution of interplanetary dust and reveal the cosmic infrared background radiation.

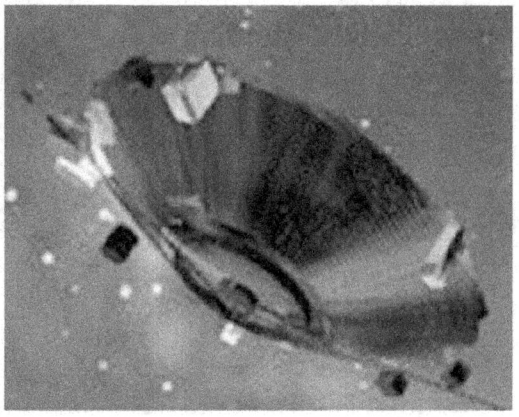

This great journey requires advanced propulsion, and the 200-kg Interstellar Probe is designed to use a 200-m-radius solar sail to achieve a velocity of 14 AU/yr. After exiting the heliosphere within a decade of launch, it will be capable of continuing to ~400 AU. Interstellar Probe will serve as the first step in a more ambitious program to explore the outer solar system and nearby galactic neighborhood.

Scientific Goals:

- Explore the nature of the interstellar medium and its implications for the origin and evolution of matter in the galaxy and the universe
- Explore the influence of the interstellar medium on the solar system, its dynamics, and its evolution
- Explore the impact of the solar system on the interstellar medium as an example of the interaction of a stellar system with its environment
- Explore the outer solar system in search of clues to its origin, and to the nature of other planetary systems

GSFC Role: TBD

Status: Conceptual development

Stellar Imager (SI)

Background: SI is a UV/optical deep-space telescope to image stars like the Sun with 0.1 milli-arcsec resolution, to help understand the solar dynamo, the internal structure and dynamics of magnetically active stars, how magnetic activity drives space weather on timescales of years to billions of years, and how magnetic activity affects planetary climates and habitability.

Scientific Goals:
Develop and test a predictive dynamo model for the Sun by:
- Observing the patterns in surface magnetic activity for a large sample of Sun-like stars (with ~1000 resolution elements on surfaces of nearby stars)
- Imaging the structure and differential rotation of stellar interiors via asteroseismology with over 30 resolution elements on stellar disks
- Carrying out a population study of Sun-like stars to determine the dependence of dynamo action on mass, internal structure, flow patterns, and time. This will enable testing of dynamo models over a few years of observations of many stars, instead of over many decades using only the Sun.

GSFC Role: TBD

Status: Conceptual development

APPENDIX 5: ACRONYM LIST

AARP	American Association of Retired Persons
AAS	American Astronomical Society
ACE	Advanced Composition Explorer
ACES	Auroral Current and Electrodynamics Structure
ACRIM	Active Cavity Radiometer Irradiance Monitor
AFRL	Air Force Research Laboratory
AGU	American Geophysical Union
AIA	Atmospheric Imaging Assembly
AIM	Aeronomy of Ice in the Mesosphere
AIP	Astrophysical Institute of Potsdam
ALI-ARMS	Accelerated Lambda Iterations for Atmospheric Radiation and Molecular Structure
AMS	American Meteorological Society
ASI	All Sky Imagers
ASIC	Application-Specific Integrated Circuit
ASTID	Astrobiology Instrument Development
ATC	Advanced Technology Center
AU	Astronomical Unit, the Earth–Sun distance, $\sim 1.5 \times 10^6$ km
BP	Bright Point (Coronal)
CAPS	Cassini Plasma Spectrometer
CAWSES	Climate and Weather of the Sun–Earth System
CBC	Canadian Broadcasting Corporation
CCD	Charged Coupled Device
CCMC	Community Coordinate Modeling Center
CDAP	Cassini Data Analysis Program
CDAWeb	Coordinated Data Analysis Web
CDF	Common Data Format
CDS	Coronal Diagnostic Spectrometer
CETP	Centre de Recherches en Physique de l'Environment Terrestre et Planetaire (France)
CFC	Combined Federal Campaign
CINDI	Coupled Ion Neutral Dynamic Investigation
CIPS	Cloud Imaging and Particle Size Experiment
CME	Coronal Mass Ejection
C/NOFS	Communications/Navigation Outage Forecasting System
CNRS	Centre National de Recherche Scientifique (France)
COHOWeb	Coordinated Heliospheric Observations Web
Co-I	Co-Investigator
COR1	Inner coronagraph on STEREO SECCHI
COSPAR	Committee on Space Research
CRCM	Comprehensive Ring Current Model
CRIS	Cosmic Ray Isotope Spectrometer

CRS	Cosmic Ray Subsystem
DC	Direct Current
DDSC	Deputy Director's Council of Science
DEM	Differential Emission Measure
DES	Dual Electron Spectrometer
DSCOVR	Deep Space Climate Observatory
DL	Double Layer
DLN	Distance Learning
DoD	Department of Defense (US)
DSN	Deep Space Network
ECCS	European Cooperation for Space Standardization
EIS	Extreme-ultraviolet (EUV) Imaging Spectrometer
EIT	EUV Imaging Telescope (on SOHO)
ENA	Energetic Neutral Atoms
EOF	Experimenters' Operations Facility
EPACT	Energetic Particles Acceleration, Composition, and Transport
E/PO	Education and Public Outreach
EPOESS	Education and Public Outreach for Earth and Space Science
EPRI	Electrical Power Research Institute
ESA	European Space Agency
ESF	European Spread F
EUNIS	Extreme-Ultraviolet Normal-Incidence Spectrograph
EUV	Extreme Ultraviolet
EUVE	Extreme Ultraviolet Explorer
EVE	Extreme ultraviolet Variability Experiment
FAC	Field Aligned Current
FAST	Fast Auroral Snapshot Explorer
FASTSAT	Fast, Affordable, Science and Technology Satellite
FGM	Fluxgate Magnetometer
FIP	First Ionization Potential
FITS	Flexible Image Transport System
FPI	Fast Plasma Instrument
FRBR	Functional Requirements for Bibliographic Data
FTE	Full-Time Equivalent
FTE	Flux Transfer Event
FTS	Fourier Transform Spectrometer
FUV	Far-Ultraviolet
FYS	First Year Seminar
GBO	Ground-based Observatory
GEC	Geospace Electrodynamics Connections
GEM	Geospace Environment Modeling
GEST	Goddard Earth Science and Technology

GGS	Global Geospace Science
GIC	Geomagnetically Induced Current
GLAST	Gamma Ray Large Area Space Telescope (former name of Fermi)
GME	Goddard Medium Energy Experiment
GMU	George Mason University
GOES	Geostationary Operational Environmental Satellite
GPS	Global Positioning System
GRL	Geophysical Research Letters
GSG	Global Scenario Group
GSRP	Graduate Student Researchers Program
GUVI	Global Ultraviolet Imager
HDMC	Heliophysics Data and Modeling Consortium
HELEX	Heliophysical Explorers
HELM	Heliophysics Event List Manager
HGO	Heliophysics Great Observatory
HI	Heliospheric Imager (on STEREO)
HMI	Helioseismic and Magnetic Imager (on SDO)
HSD	Heliophysics Science Division
HST	Hubble Space Telescope
IBEX	Interstellar Boundary Explorer
ICESTAR	Interhemispheric Conjugacy Effects in Solar Terrestrial and Aeronomy Research
IDL	Interactive Data Language
IGY	International Geophysical Year
IHY	International Heliophysical Year
IMACS	Imaging Spectrograph of Coronal electrons
IMAGE	Imager for Magnetopause-to-Aurora Global Exploration
IMP	Interplanetary Monitoring Platform
IMPACT	*In situ* Measurements of Particles and CME Transients
IMS	Ion Mass Spectrometer
INMS	Ion-Neutral Mass Spectrometer
IPY	International Polar Year
IR	Infrared
IRAD	Independent Research and Development
IRI	International Reference Ionosphere
ISAS	Japan's Institute for Space and Aeronautical Science
ISCORE	Imaging Spectrograph of Coronal Electrons
ISEE	International Sun Earth Explorer
ISIS	International Satellites for Ionospheric Studies
ISO	International Standards Organization
ISTP	International Solar Terrestrial Physics
IT	Ionosphere-Thermosphere [region]
ITSP	Ionosphere-Thermosphere Storm Probes
IUE	Internal Ultraviolet Explorer

JAXA	Japan Aerospace Exploration Agency
JGR	Journal of Geophysical Research
LADEE	Lunar Exosphere and Dust Environment Explorer
LADTAG	Lunar Airborne Dust Toxicity Advisory Group
LASCO	Large Angle and Spectrometric Coronagraph (on SOHO)
LENA	Low Energy Neutral Atom
LEO	Low Earth Orbit
LLBL	Low-Latitude Boundary Layer
LM	Lockheed Martin
LOS	Line-of-Sight
LRO	Lunar Reconnaissance Orbiter
LTE	Local Thermodynamic Equilibrium
LW	Long Wavelength
LWS	Living With a Star
MACS	Goddard's Multi-Aperture Coronal Spectrograph
MAG	Magnetometer
MagCon	Magnetospheric Constellation
MC	Magnetic Cloud
MCAT	Magnetic Cloud Analysis Tool
MDI	Michelson Doppler Imager (on SOHO)
MDR	Mission Definition Review
MENA	Medium-Energy Neutral Atom
MESSENGER	Mercury Surface, Space Environment, Geochemistry and Ranging
MFI	Magnetic Field Investigation
MGS	Mars Global Surveyor
MHD	Magnetohydrodynamic
MINI-ME	Miniature Imager Neutral Ionospheric atoms and Magnetospheric
MLSO	Mauna Loa Solar Observatory
MLTI	Magnetosphere mission concept
MMC	Magnetosphere Mission Concept
MMS	Magnetospheric MultiScale
MO&DA	Mission Operations and Data Analysis
MOC	Mission Operations Center
MOR	Mission Operations Room
MOSES	Multi-Order Solar EUV Spectrograph
NAC	NASA Advisory Council
NESC	NASA Engineering and Safety Center
NOAA	National Oceanographic and Atmospheric Administration
NPP	NASA Postdoctoral Program
NRC	National Research Council
NRL	National Research Laboratory

NSF	National Science Foundation
NSSDC	National Space Science Data Center
OPR	Outer Planets Research
PDR	Preliminary Design Review
PEACE	Plasma Electron and Current Experiment
PI	Principal Investigator
PIC	Particle-in-Cell
PICARD	Not an acronym, but the name of a mission.
PISA	Plasma Impedance Spectrum Analyzer
PMC	Polar Mesospheric Clouds
PSD	Planetary Science Division
PWG	Polar-Wind-Geotail
R&D	Research and Development
RAISE	Rapid Acquisition Imaging Spectrograph Experiment
RBE	Radiation Belt Environment
RBSP	Radiation Belt Storm Probes
RHESSI	Ramaty High Energy Solar Spectroscopic Imager
RPI	Radio Plasma Imager
RPWS	Radio and Plasma Wave Science
SABER	Sounding of the Atmosphere using Broadband Emission Radiometry
SBIR	Small Business Innovative Research
SCIFER	Sounding of the Cleft Ion Fountain Energization Region
SCOSTEP	Scientific Committee on Solar-Terrestrial Physics
SDAC	Solar Data Analysis Center
SDAT	Science Data Analysis Tool
SDO	Solar Dynamics Observatory
SECCHI	Sun–Earth Connection Coronal and Heliospheric Imager
SECEF	Sun–Earth Connection Education Forum
SEE	Solar EUV Experiment
SERB	Space Experiment Review Boards
SERTS	Solar Extreme-ultraviolet Research Telescope and Spectrograph
SESDA	Space and Earth Science Data Analysis
SHINE	Solar, Heliospheric, and Interplanetary Environment
SI	Stellar Imager
SIM	Spectral Irradiance Monitor
SIS	Science Information Systems
SIS	Solar Isotope Spectrometer
SMD	Science Mission Directorate
SMEX	Small Explorer
SOHO	Solar and Heliospheric Observatory
SOLSTICE	Solar Stellar Irradiance Comparison Measurement

SORCE	Solar Radiation and Climate Experiment
SOT	Solar Optical Telescope (on Hinode)
SP+	Solar Probe Plus
SPASE	Space Physics Archive Search and Extract
SPD	Solar Physics Division of the AAS
SPDF	Space Physics Data Facility
SPSO	Science Proposal Support Office
SSAC	Space Science Advisory Committee
SSC	STEREO Science Center
SSMO	Space Science Mission Operations
SSREK	Solar System Radio Explorer Kiosk
ST5	Space Technology 5
STDT	Science and Technology Definition Team
STEREO	Solar Terrestrial Relations Observatory
STP	Solar Terrestrial Probe
SUMI	Solar Ultraviolet Normal Magnetograph Investigation
SVS	Scientific Visualizations Studio
SW	Short Wavelength
SWAN	Space Weather Awareness at NASA
SWAVES	Not an acronym, but a series of investigations on the STEREO spacecraft (see WAVES).
SWE	Solar Wind Experiment
SWICS	Solar Wind Ionic Composition Spectrometer
SWMF	Space Weather Modeling Framework
SXT	Soft X-Ray Telescope (Yohkoh)
TECHS	Thermal Electron Capped Hemisphere Spectrometer
TES	Thermal Emission Spectrometer
TGF	Terrestrial Gamma-ray Flashes
THEMIS	Time History of Events and Macroscale Interactions during Substorms
TIDE	Thermal Ion Dynamics Experiment
TIGER	Trans-Ion Galactic Element Recorder
TIM	Total Irradiance Monitor
TIMS	Technical Information and Management Services
TIMED	Thermosphere Ionosphere Mesosphere Energetics and Dynamics
TOPIST	Topside Ionospheric Scaler with True height (algorithm)
TRACE	Transition Region and Coronal Explorer
TRICE	Twin Rockets to Investigate Cusp Electrodynamics
TSSM	Titan Saturn System Mission
TTI	Thermospheric Temperature Imager
TR&T	Targeted Research and Technology
TWINS	Two Wide-Angle Imaging Neutral-Atom Spectrometers
ULF	Ultra-Low Frequency

UNBSS	United Nations Basic Space Science
USAF	United States Air Force
USNA	United States Naval Academy
USRA	Universities Space Research Association
UV	Ultraviolet
UVSC	UV Spectro-Coronagraph
VDF	Velocity Distribution Function
VEFI	Vector Electric Field Instrument
VEPO	Virtual Energetic Particles Observatory
VERIS	Very High Angular Resolution Imaging Spectrometer
VGA	Venus Gravity Assist
VHO	Virtual Heliospheric Observatory
VIM	Voyager Interstellar Mission
VIRGO	Variability of Solar Irradiance and Gravity Oscillations
VIS	Visible Imaging System
VITMO	Virtual Ionospheric/Thermospheric/Mesospheric Observatory
VLF	Very Low Frequency
VMO/G	Virtual Magnetic Observatory at Goddard
VRML	Virtual Reality Modeling Language
VSO	Virtual Solar Observatory
VWO	Virtual Wave Observatory
VxO	Virtual discipline Observatory
WAVES	Not an acronym, but a series of investigations on the Wind spacecraft that will provide comprehensive coverage of radio and plasma wave phenomena.
Wind	Not an acronym, but a NASA spacecraft in the Global Geospace Science Program
XMM	X-ray Multi-Mirror Mission
XRT	X-Ray Telescope

www.ingramcontent.com/pod-product-compliance
Lightning Source LLC
Chambersburg PA
CBHW081723170526
45167CB00009B/3682